Machine Shop Operations

VISUTEXT

S. F. Krar
Technical Director
Welland Eastdale Secondary School
Welland, Ontario

J. W. Oswald
Technical Director
Westlane Secondary School
Niagara Falls, Ontario

J. E. St. Amand
Formerly Technical Coordinator
West Haldimand District High Schools
Cayuga, Ontario

McGraw-Hill Ryerson Limited

Toronto Montreal New York London Sydney
Johannesburg Mexico Panama Düsseldorf Singapore
Sao Paulo Kuala Lumpur New Delhi

ISBN 0-07-082200-X

1 2 3 4 5 6 7 8 9 10 AP 4 3 2 1 9 8 7 6 5

Printed and bound in Canada

The authors wish to express their sincere thanks and appreciation to Alice H. Krar for her untiring devotion in reading, typing, and checking the manuscript for this text. Without her efforts this text could not have been produced.

Table of Contents

The English system of measurement is common in Canada and the United States. However, the Metric system, which is used by over 80 per cent of the countries in the world, is being gradually introduced since it is not as complicated as the English system.

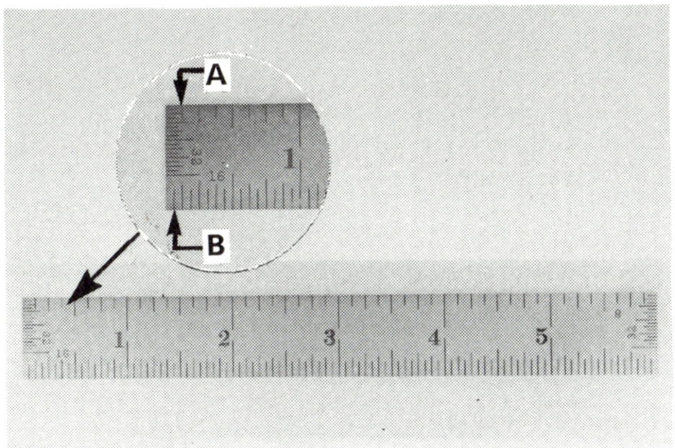

1. **ENGLISH SYSTEM (RULES)**
   - A rule is used for measurements that do not require an accuracy greater than 1/64" (.39 mm).
   - The most common rule is the one with #4 graduations.
   - It is graduated in 8ths and 16ths on one side.
   - The 8th graduations (A) mean that the inch is divided into eight equal parts.
   - The 16th graduations (B) mean that the inch is divided into sixteen equal parts.

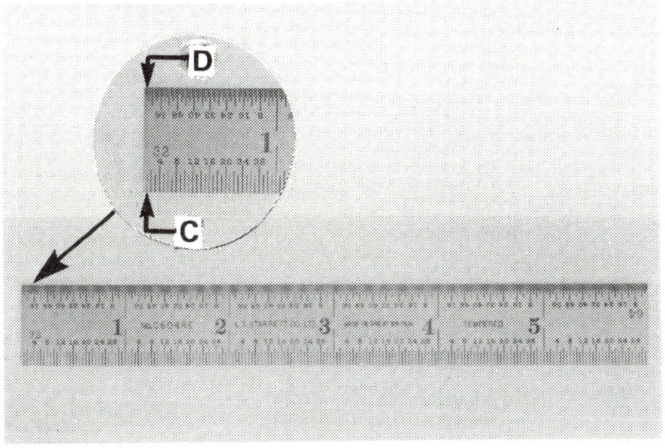

2.
   - The other side of a rule with #4 graduations is graduated in 32nds and 64ths.
   - The 32nd graduations (C) mean that the inch is divided into thirty-two equal parts. Every fourth line is numbered for easy reading.
   - The 64th graduations (D) mean that the inch is divided into sixty-four equal parts. Every eighth line is numbered for easy reading.

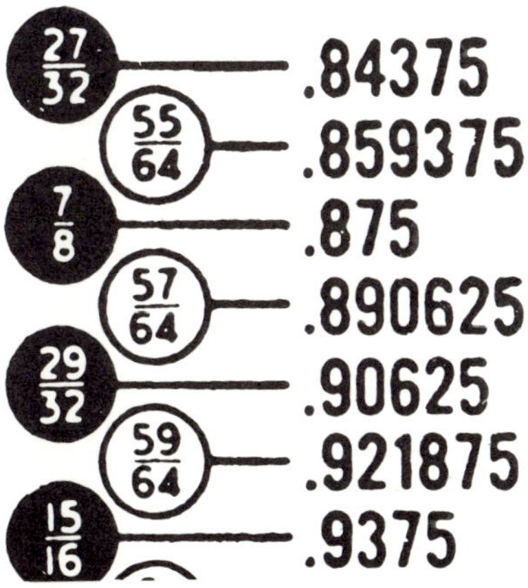

7/8 = 7.000 / 8

```
    .875
8 / 7.000
    6 4
    ---
     60
     56
     --
      40
```

3. **ENGLISH SYSTEM (DECIMALS)**

   The most common decimal fraction is thousandths of an inch (three numbers to the right of the decimal point). To convert a fraction to decimals:
   (1) Place a decimal point (.) to the right of the last number of the numerator (top number).
   (2) Add three zeros (000) after the decimal point.
   (3) Divide the numerator by the denominator (bottom number).

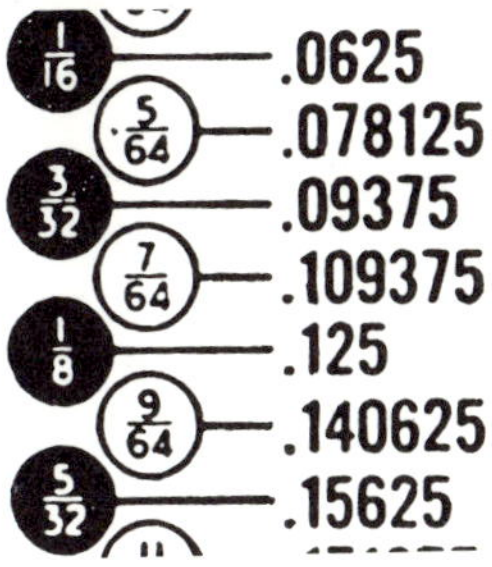

3/32 = 3.000 / 32

```
  E
   ↘.0937 or .094
32 / 3.0000
     288
     ---
      120
       96
      ---
      240
      224
```

4. To change 3/32" to decimals (thousandths of an inch).
   (1) Place a decimal point (E) immediately above the decimal in 3.000.
   (2) Since 32 will not divide into 30, a zero (0) must be placed to the right of the decimal.
   (3) Divide 32 into 3000 which will go 9 times.
   (4) Divide to four places in order to round out the decimal number to .094.

.375

hundreds tens units

5. (1) The first number to the right of the decimal represents hundreds (100 to 900) $3 \times 100 = 300$
   (2) The second number represents tens (10 to 90) $7 \times 10 = 70$
   (3) The third number represents units (1 to 10) $5 \times 1 = 5$

   $\overline{.375}$

   .375 would mean three hundred and seventy-five thousandths of an inch.

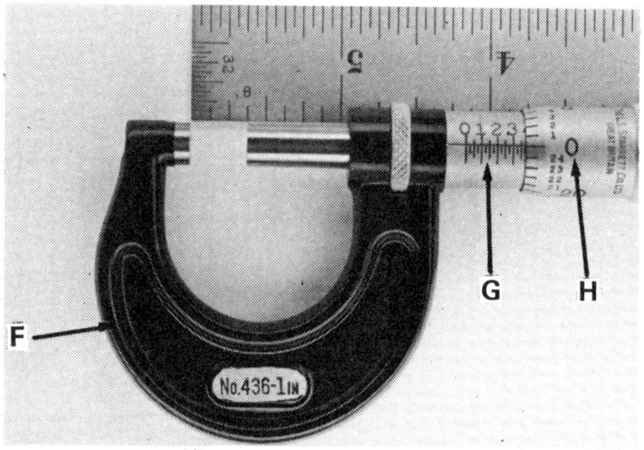

6. The standard micrometer (F), which measures in thousandths of an inch, is the most common decimal measuring instrument.
   - Each line on the barrel (G) equals .025
   - Each numbered line on the barrel equals .100
   - Each line on the thimble (H) has a value of .001

   Micrometer Reading

The #3 on the barrel	$= 3 \times .100$ or	.300
Three lines past #3	$= 3 \times .025$ or	.075
	Total Reading	.375

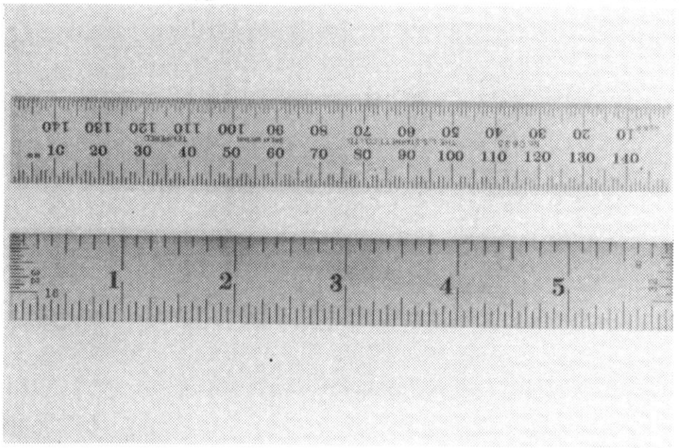

7. METRIC SYSTEM (RULES)
   The most common metric rules are graduated in millimeters (mm) and half millimeters (½ mm). A comparison of English and Metric rule graduations follows:

English		Metric
1/16″	=	1.58 mm
⅛″	=	3.17 mm
¼″	=	6.35 mm
½″	=	12.70 mm
1″	=	25.40 mm

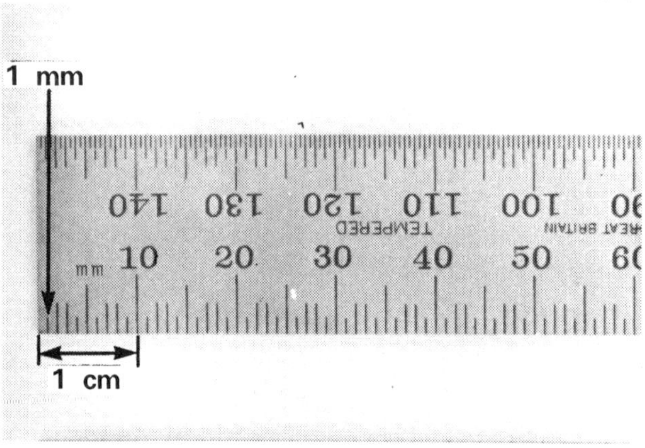

8. The meter is the standard unit of the metric system. All metric units are larger or smaller than the meter by multiples of ten.

Milli (mm) 1/1000 of a meter	(10 mm = 1 cm)
Centi (cm) 1/100 of a meter	(10 cm = 1 dm)
Deci (dm) 1/10 of a meter	(10 dm = 1 m)
Meter (m) basic unit	
Deka (dkm) 10 meters	(10 dkm = 1 hm)
Hecto (hm) 100 meters	(10 hm = 1 km)
Kilo (km) 1000 meters	

9.

Linear Units	METRIC SYSTEM Abbreviation	Value	ENGLISH SYSTEM Value
1 meter	M or m	1. meter	39.37 inches
1 decimeter	dm	.1 meter	3.937 inches
1 centimeter	cm	.01 meter	.394 inches
1 millimeter	mm	.001 meter	.039 inches
1 dekameter	dkm	10. meter	32.80 feet
1 hectometer	hm	100. meter	328.00 feet
1 kilometer	km	1000. meter	.6214 mile

CONVERSION OF INCHES TO MILLIMETERS

Inches	Milli-meters	Inches	Milli-meters	Inches	Milli-meters
.001	.025	.290	7.37	.660	16.76
.002	.051	.300	7.62	.670	17.02
.003	.076	.310	7.87	.680	17.27
.004	.102	.320	8.13	.690	17.53
.005	.127	.330	8.38	.700	17.78
.006	.152	.340	8.64	.710	18.03
.007	.178	.350	8.89	.720	18.29
.008	.203	.360	9.14	.730	18.54
.009	.229	.370	9.40	.740	18.80
.010	.254	.380	9.65	.750	19.05

10. **CONVERSION OF MILLIMETERS TO INCHES**

Milli-meters	Inches	Milli-meters	Inches	Milli-meters	Inches
.01	.0004	.35	.0138	.68	.0268
.02	.0008	.36	.0142	.69	.0272
.03	.0012	.37	.0146	.70	.0276
.04	.0016	.38	.0150	.71	.0280
.05	.0020	.39	.0154	.72	.0283
.06	.0024	.40	.0157	.73	.0287
.07	.0028	.41	.0161	.74	.0291
.08	.0031	.42	.0165	.75	.0295
.09	.0035	.43	.0169	.76	.0299
.10	.0039	.44	.0173	.77	.0303

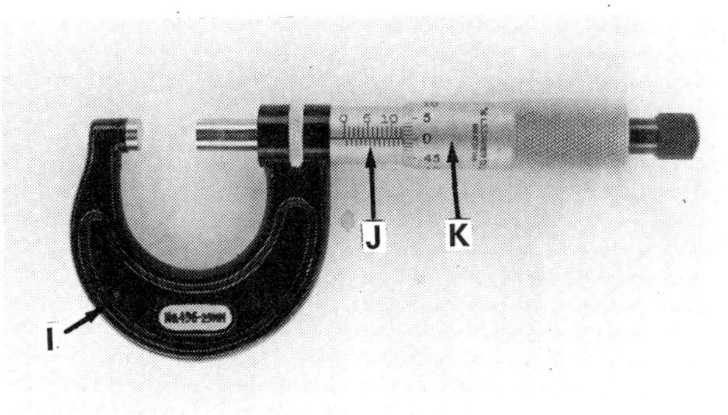

11. The metric micrometer (I), which measures in .01 mm (one-hundredth of a millimeter), is the most common instrument used for precision measurement.
- Each line on the barrel (J) equals .50 mm
- Each numbered line on the barrel equals 5. mm
- Each line on the thimble (K) equals .01 mm

Micrometer Reading
The #10 on the barrel equals 10×1 or 10. mm
Five lines past #10 equals 5×.50 or 2.50 mm
Total Reading 12.50 mm

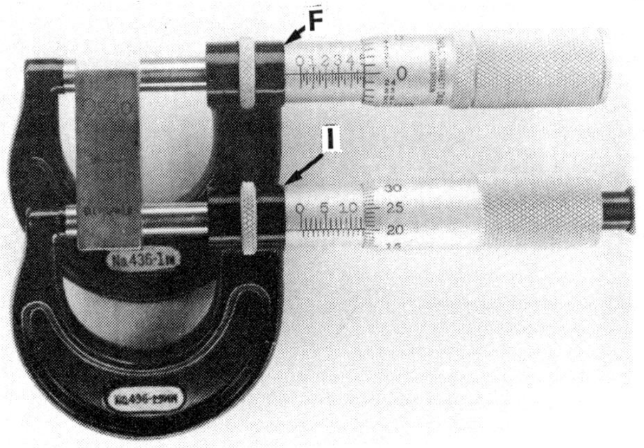

12. Figure 12 shows an English and Metric micrometer measuring the same piece.
- The English micrometer (F) shows a measurement of .500
- The Metric micrometer (I) shows a measurement of 12.70 mm

The steel rule, manufactured in a variety of types and sizes, is probably the most commonly used measuring tool in a machine shop. Steel rules should not be used where accuracy of more than 1/64" (.39 mm) is required.

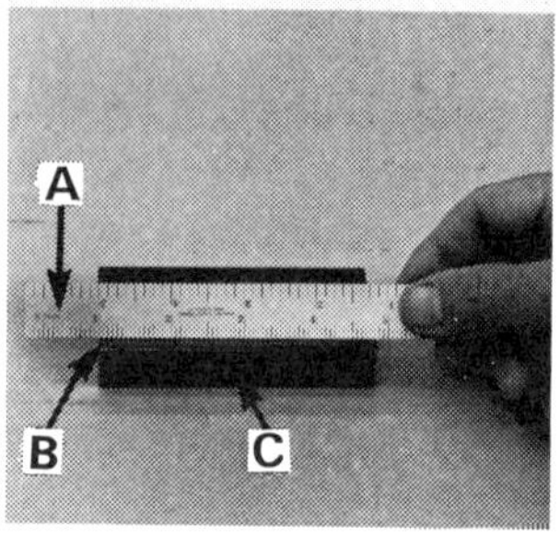

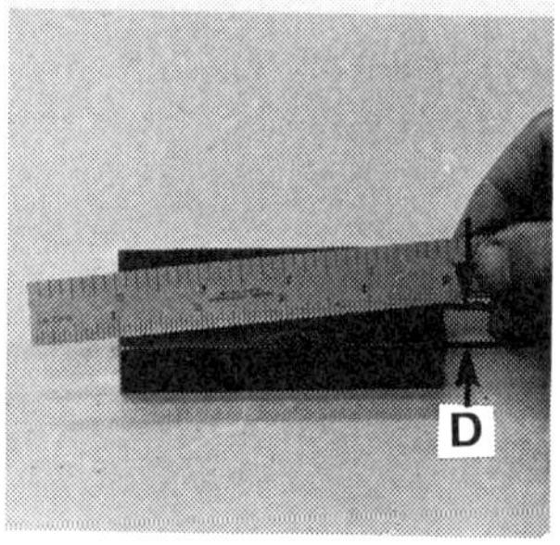

1. **MEASURING LENGTH**
   - Position the rule (A) so that the center of the 1" line is on the edge (B) of the workpiece.
   - Hold the rule parallel to the edge (C) of the workpiece.
   - When the rule is not held parallel (D) the measurement will be incorrect. Note the difference in the measurements of the same workpiece in the two photos.

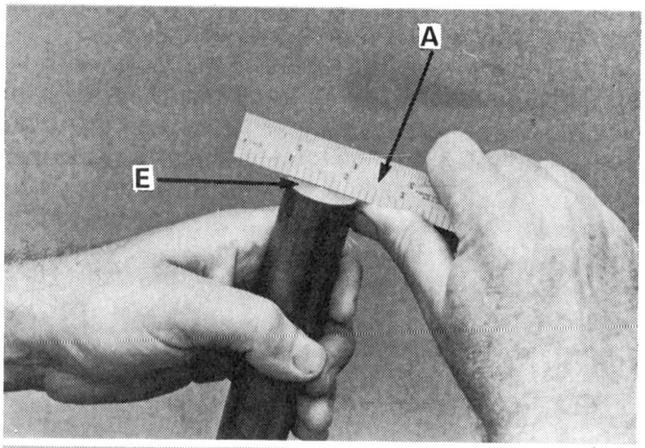

2. **MEASURING ROUND WORK**
   - Remove the burrs from the edge of the workpiece (E) with a file.
   - Hold the work in one hand.
   - Hold the rule (A) in the other hand and position a main graduation line on the edge of the work.
   - Position the rule over the center line of the workpiece.
   - Press the thumb against the work to steady the rule.
   - Note the reading on the rule.

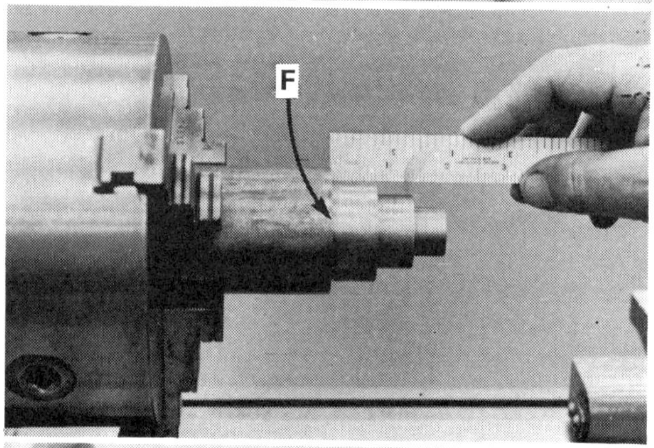

3. **MEASURING TO A SHOULDER**
   - Place the end of the rule against the shoulder (F). It is neither convenient nor accurate enough to measure from a graduation on the rule.
   - Hold the rule parallel to the edge or the center line of the workpiece.
   - Note the reading on the rule.

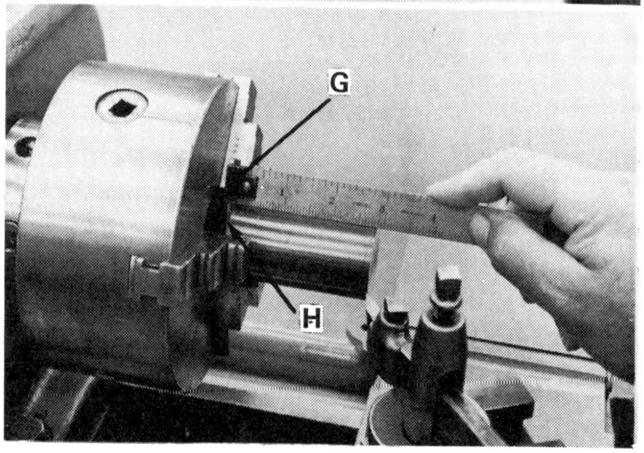

4. **MEASURING WITH A HOOK RULE**

   A hook rule is used when it is not convenient to measure from the end or from a graduation on a rule.
   - Place the hook (G) over the edge (H) or shoulder of the work.
   - Hold the rule parallel to the edge or the center line of the work.
   - Note the reading on the rule.

EXTERNAL MEASUREMENT

External measurements may be made with a rule, outside caliper, micrometer, or vernier caliper. The instrument used will depend upon the accuracy required.

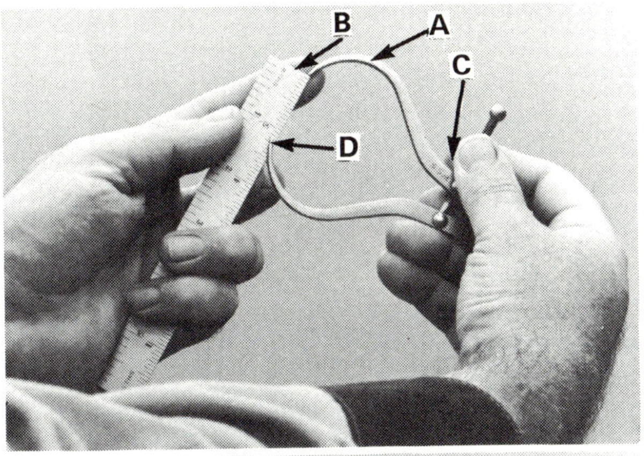

1. **OUTSIDE CALIPER AND RULE—for measurements up to 1/64" (.39 mm) accuracy.**
   - Place one leg of the caliper (A) over the end of the rule (B).
   - Keep the caliper leg in position with a finger as shown.
   - Hold both legs of the caliper parallel to the edge of the rule.
   - Turn the adjusting nut (C) with the thumb and forefinger until the end of the caliper leg (D) splits the desired graduation on the rule.

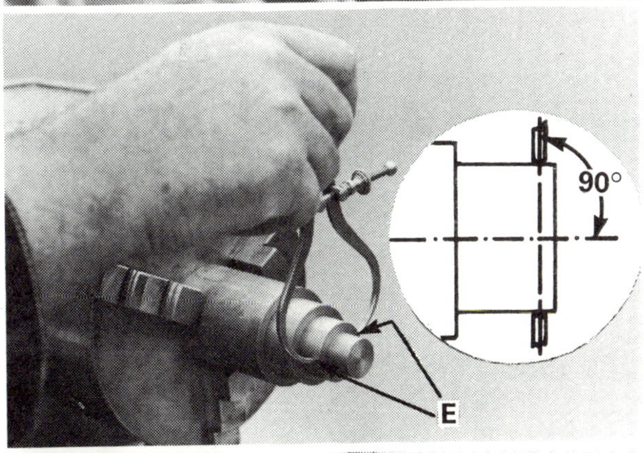

2.
   - Place the caliper on the work so that the line between the ends of the caliper legs (E) is at right angles to the center line of the work.
   - Hold the caliper lightly between the thumb and forefinger.

     NOTE: If the diameter is correct, the caliper should just slip over the work by its own weight. NEVER force a caliper over a diameter.

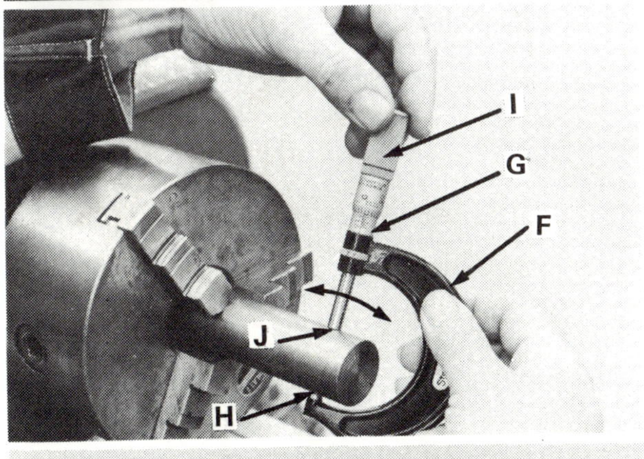

3. **MICROMETER CALIPERS (.001 accuracy)**
   - Hold the micrometer frame (F) in the right hand, with the graduations on the barrel (G) facing the operator.
   - Place the anvil (H) on the underside of the work.
   - Turn the thimble (I) and rock the micrometer slightly (in the direction of the arrows) until the spindle (J) contacts the work.
   - Use the ratchet stop or friction thimble to adjust the micrometer for the proper tension.
   - Note the micrometer reading.

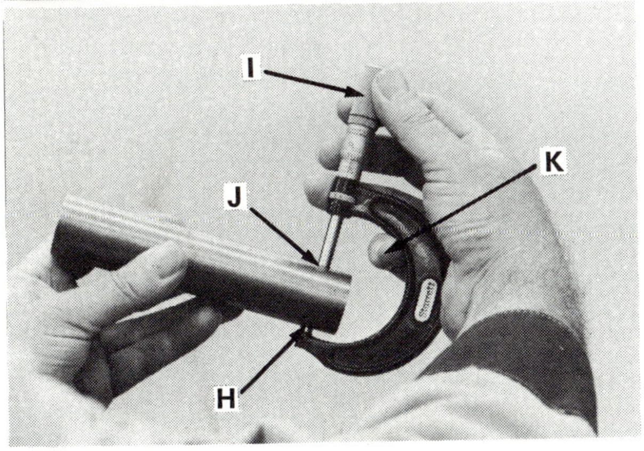

4. **MEASURING SMALL WORKPIECES**
   - Hold the micrometer with the small finger (K) around the frame and the thumb and forefinger holding the thimble (I).
   - Place the anvil (H) on the underside of the work.
   - Turn the thimble (I) and rock the micrometer slightly until the spindle (J) contacts the work.
   - Use the ratchet stop or friction thimble to adjust the micrometer for the proper tension.
   - Note the micrometer reading.

Internal measurement may be made with inside calipers, small hole gages, telescope gages, inside micrometers, dial type calipers, and vernier calipers. The accuracy required will determine which measuring tool should be used.

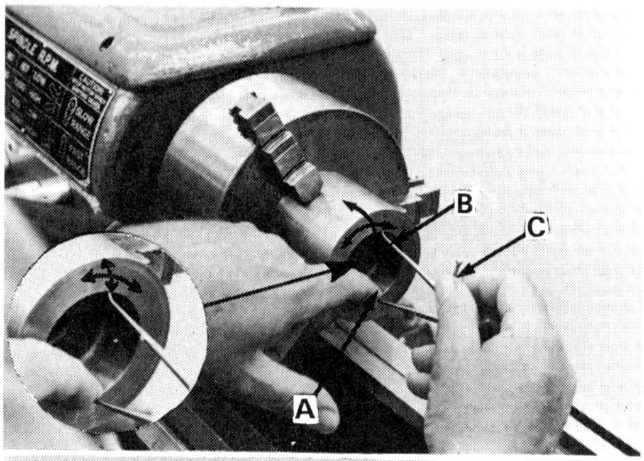

1. INSIDE CALIPERS
   - Place one leg of the caliper in the hole at the 6 o'clock position (A) and hold it with a finger.
   - Hold the caliper parallel to the edge of the hole with the upper leg (B) at the 12 o'clock position.
   - Move the top leg in the direction of the arrows and turn the adjusting nut (C) until *only a slight drag* is felt on the caliper leg.

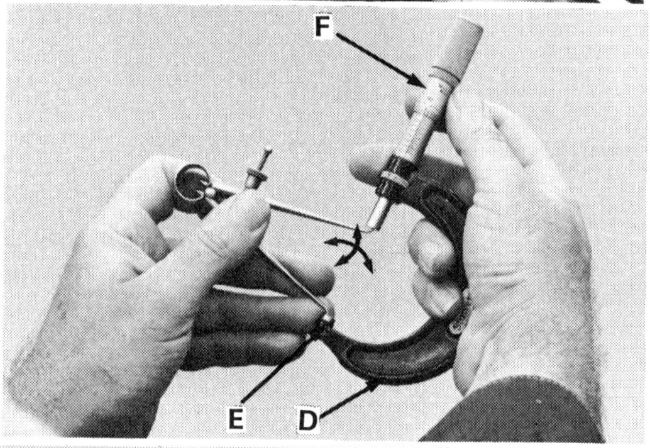

2.
   - Hold the micrometer (D) in the right hand.
   - Place one leg of the caliper on the micrometer anvil (E) and hold it in position with a finger.
   - Rock the top leg of the caliper in the direction of the arrows.
   - Adjust the micrometer thimble (F) until *only a slight drag* is felt as the caliper leg passes the measuring face.
   - Note the micrometer reading.

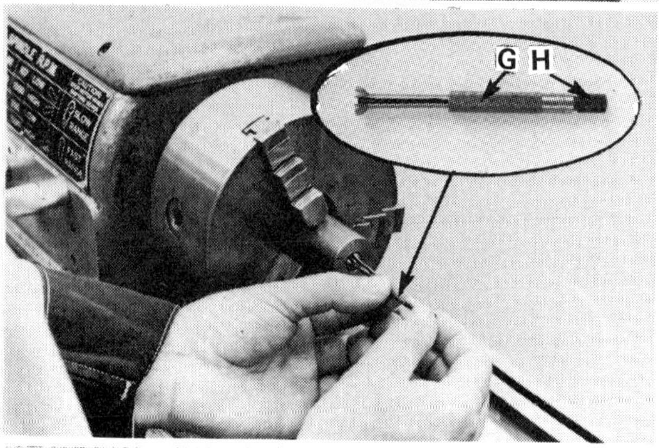

3. SMALL HOLE GAGES — for holes up to ½" (12.7 mm) diameter.
   - Hold the body of the gage (G) with the thumb and finger of one hand and place the gage inside the edge of the hole.
   - Turn the adjusting stem (H) until the contact points touch the hole.
   - Rock the end of the gage slightly and adjust the stem (H) until a slight drag is felt on the contact points.

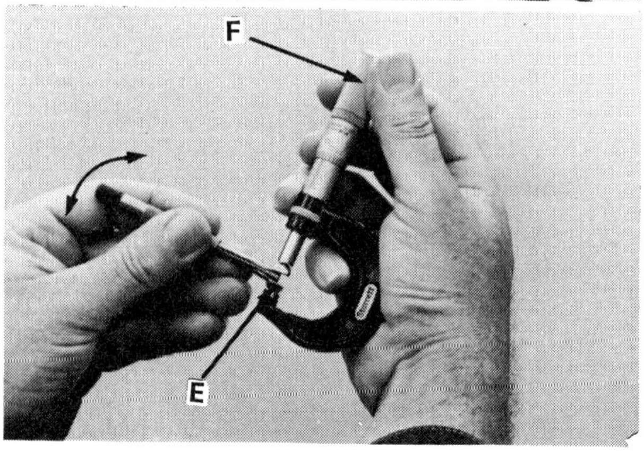

4.
   - Hold the micrometer in the right hand as shown.
   - Place the ball of the small hole gage on the micrometer anvil (E).
   - Adjust the micrometer thimble (F) until a *slight drag* is felt as the outer end of the gage is rocked slightly.
   - Note the micrometer reading.

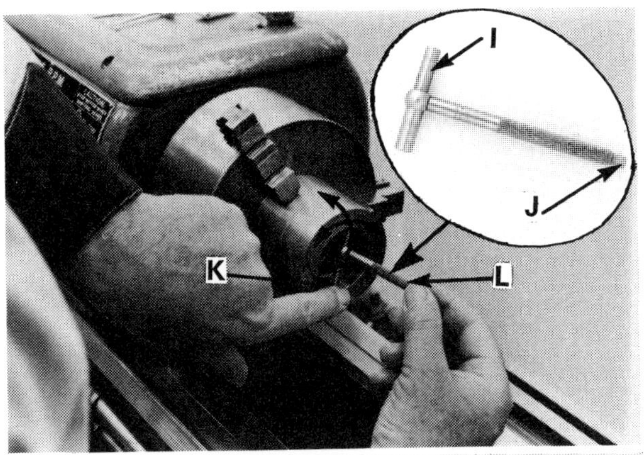

5. TELESCOPE GAGES — for holes over ½" (12.7 mm) diameter.
   - Retract the plunger(s) (I).
   - Place the gage in the hole.
   - Hold the gage vertical and loosen the knurled lock (J).
   - Hold one leg (K) in position and raise the handle (L) slightly to move the other leg past center.
   - *Snug up* the lock and lower the handle, bringing the top leg out past center.
   - *Tighten the lock* and recheck the setting.

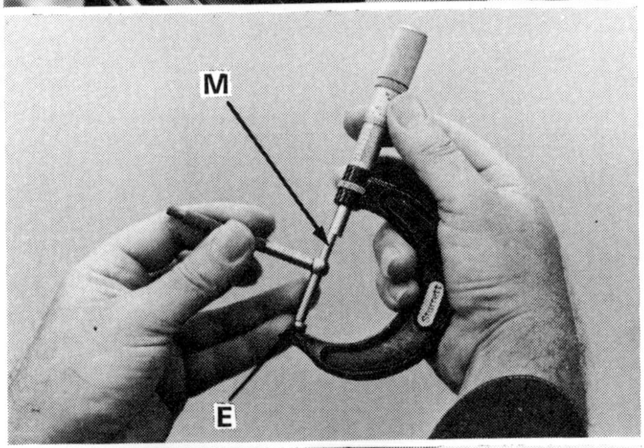

6.
   - Hold one leg of the telescope gage on the anvil (E) and support it with one finger.
   - Rock the top leg (M) of the telescope gage slightly.
   - Adjust the micrometer until the same drag is felt as when gaging the hole.
   - Note the micrometer reading.

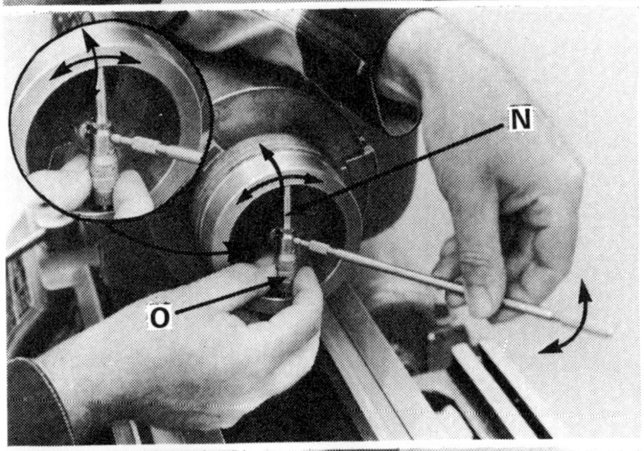

7. INSIDE MICROMETERS — for holes larger than 1½" (38.1 mm) diameter.
   - Insert the correct extension rod (N) in the micrometer for the size of the hole to be measured.
   - Place the micrometer in the hole as shown with the ends parallel to the edge of the hole.
   - Rock the top of the micrometer back and forth slightly.
   - Adjust the micrometer thimble (O) until a slight drag is felt.
   - Remove the micrometer and note the reading.

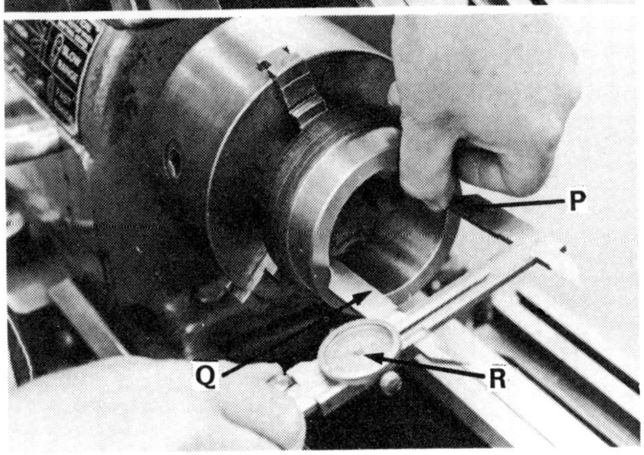

8. DIAL CALIPERS — for holes larger than ½" (12.7 mm) diameter.
   - Support the solid leg (P) of the caliper against the back edge of the hole, with one hand.
   - Adjust the movable jaw (Q) to bear against the opposite side of the hole.
   - Move the caliper up and down slightly while keeping both jaws in contact with the side of the hole.
   - Note the highest reading on the dial (R).

The combination set, consisting of the blade (rule), square head, center head and protractor head, can be used for layout and inspection. Each head is provided with a knurled nut to lock it in position on the blade.

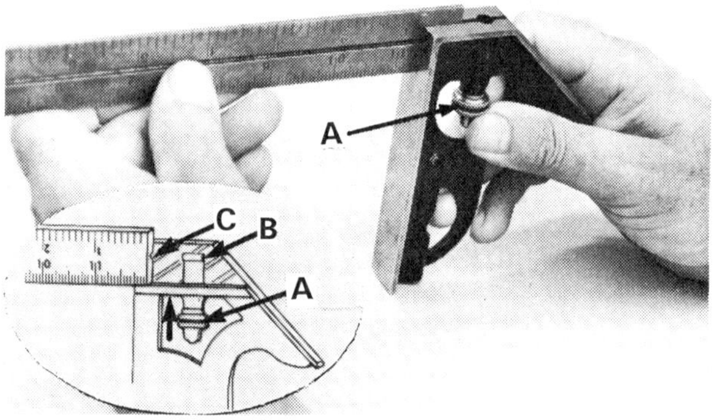

1. ASSEMBLING A HEAD AND BLADE
   - Loosen the nut (A) and press it up in the direction of the arrow.
   - Note the position of the hook (B).
   - Hold the rule (blade) so that the groove (C) lines up with the hook (B).
   - Push the rule into the head and tighten the lock nut (A).

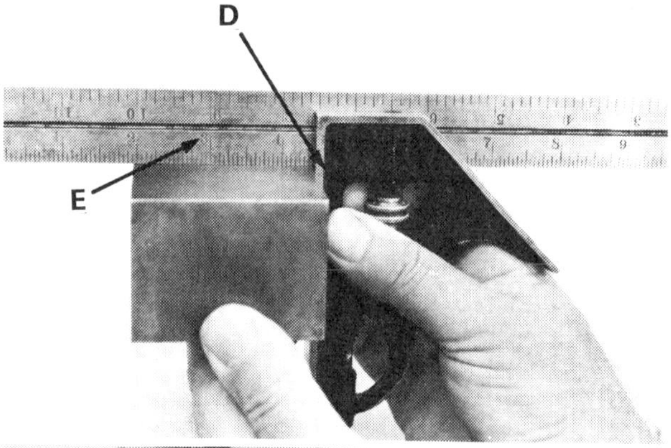

2. CHECKING FOR SQUARENESS
   - Remove the burrs from the workpiece with a file.
   - Place the body of the square (D) against a finished side of the workpiece.
   - Bring the blade (E) down lightly against the adjacent face.
   - Hold the square and workpiece up to the light.
   - If the workpiece is square, the blade will bear all along the surface and no light will be visible between the blade and the workpiece.

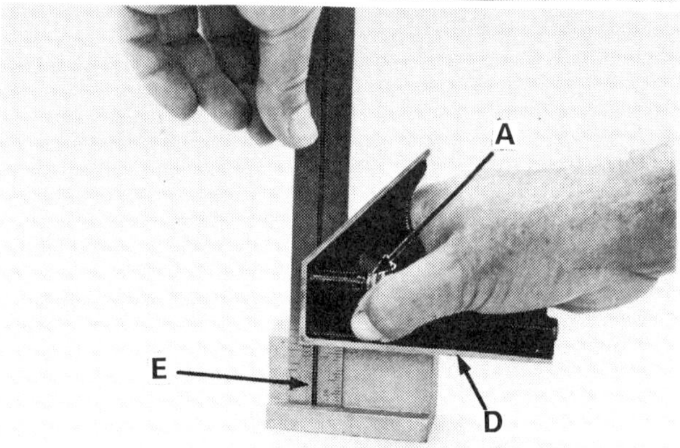

3. MEASURING DEPTHS
   - Remove the burrs from the workpiece with a file.
   - Loosen the lock nut (A) on the square.
   - Place the body of the square (D) on the upper surface.
   - Slide the blade (E) down until it contacts the bottom of the slot or step.
   - Tighten the lock nut (A) and note the reading on the rule.

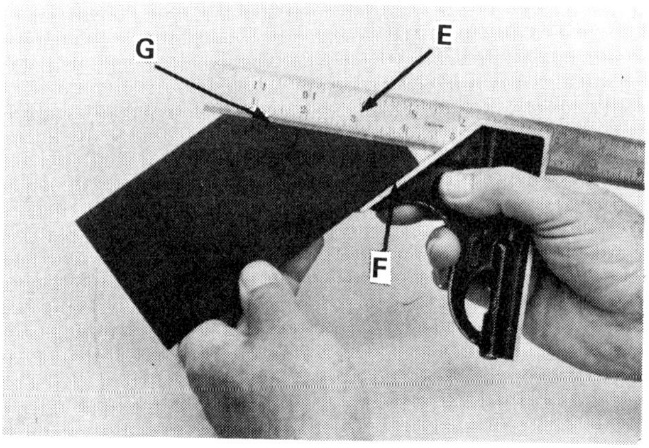

4. CHECKING A 45° ANGLE
   - Place the angled face (F) of the square on one surface of the work.
   - Slide the head along the work surface until the blade (E) contacts the angled surface (G) of the work.
   - Check that the blade bears evenly along the surface (G).

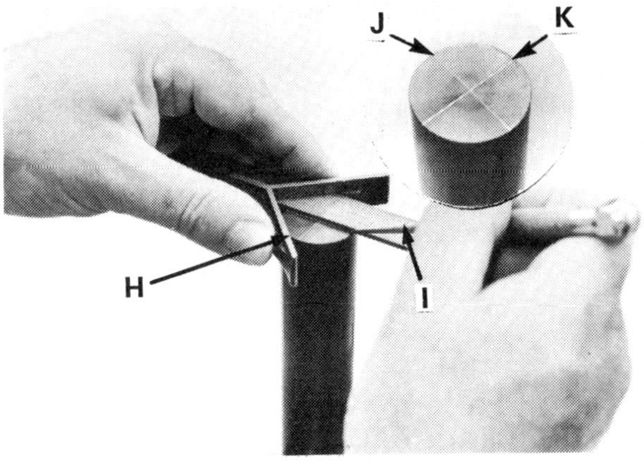

5. LAYING OUT THE CENTER OF ROUND WORK
   - Remove the burrs from the end of the workpiece.
   - Apply layout die to the end of the work.
   - Hold the center head (H) firmly against the work.
   - Hold a sharp scriber (I) at an angle so that only the point touches the edge of the blade.
   - Mark a line (J) along the edge of the blade.
   - Rotate the center head one-quarter of a turn and scribe a second line (K) at 90° to line (J).
   - Lightly center punch where the two lines cross.

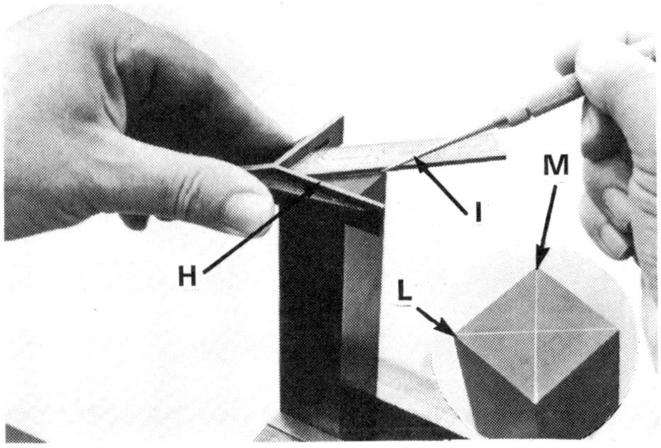

6. LAYING OUT THE CENTER OF SQUARE WORK
   - Remove the burrs from the end of the workpiece.
   - Place the center head (H) against a corner of the workpiece.
   - Hold a sharp scriber (I) on an angle and mark a line (L) along the edge of the rule.
   - Place the center head against the next corner of the work and scribe a second line (M).
   - Lightly center punch where the two lines cross.

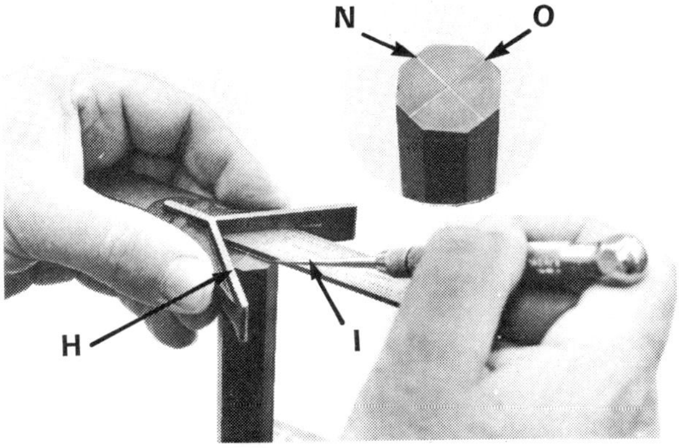

7. LAYING OUT THE CENTER OF OCTAGONAL WORK
   - Remove the burrs from the end of the workpiece.
   - Hold the center head (H) firmly against the work.
   - Hold a sharp scriber (I) on an angle and mark a line (N) along the edge of the blade.
   - Rotate the center head 90° and scribe a second line (O).
   - Lightly center punch where the two lines cross.

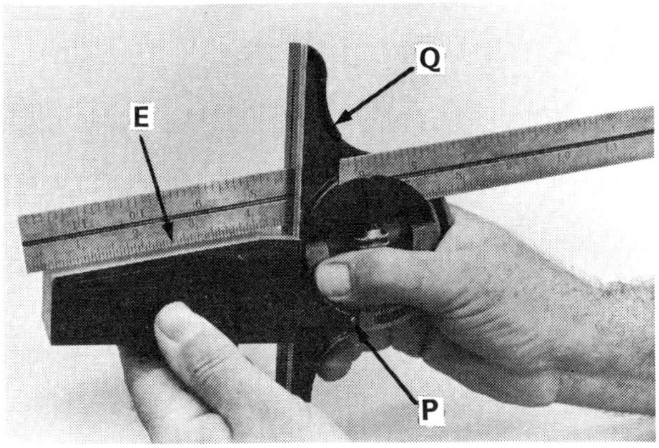

8. CHECKING AN ANGLE
   - Remove the burrs from the work.
   - Loosen the protractor locking screws (P).
   - Set the protractor head (Q) to the required angle and tighten the locking screws (P).
   - Place one edge of the work against the body of the protractor.
   - Slide the work up into contact with the blade (E).
   - Check the angle for accuracy. If the angle is correct, no light should be seen between the blade and the work surface.

THREAD MEASUREMENT

External threads may be measured with a test nut, thread micrometer, or by the three-wire method. The accuracy of the thread will determine which method should be used.

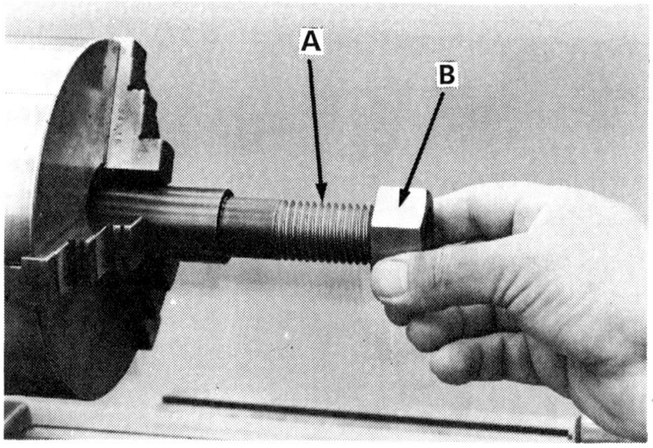

1. **TEST NUT — used when accuracy of the thread is not critical.**
   - **Remove the burrs from the top of the thread (A) with a file. The diameter of the threaded section should be .002 (.051 mm) smaller than the major diameter.**
   - **Start the test nut (B) on the thread by hand.**
   - **A correct thread requires slight pressure to turn the nut on with no end play.**

   CAUTION: Never force a nut on with a wrench.

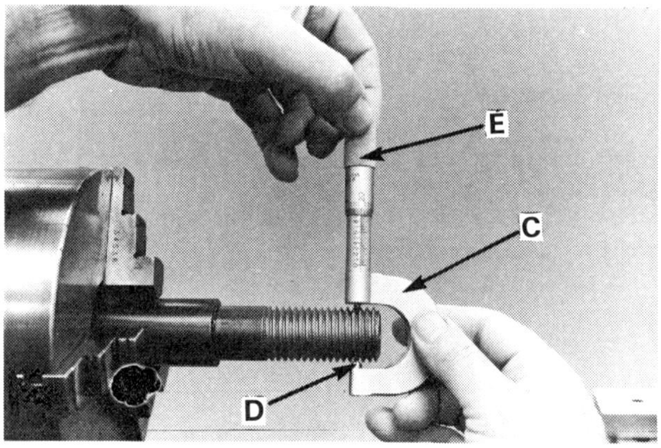

2. **SCREW THREAD MICROMETER — used to measure the pitch diameter of a thread.**
   - **Select the proper thread micrometer (C) for the pitch of thread to be measured.**
   - **Calculate the pitch diameter of the thread (major diameter minus one thread depth).**
   - **Place the swivel anvil (D) of the micrometer on the thread.**
   - **Adjust the thimble (E) until the spindle bears against the thread.**
   - **Compare the micrometer reading with the calculated pitch diameter.**

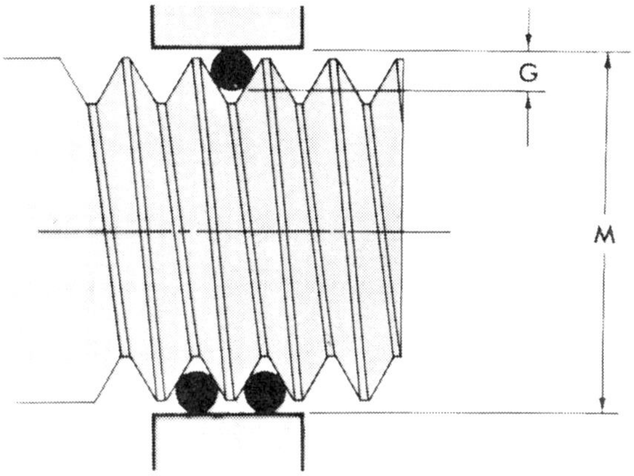

3. **THREE-WIRE METHOD — the most accurate method of measuring a thread.**

   For American 60° threads, two measurements must be calculated:

   (1) *the best wire size* **(G)**

   $$G = \frac{.57735}{N} \quad (N = \text{threads per inch})$$

   (2) *the measurement over the wires* **(M)**

   $$M = D + 3G - \frac{1.5155}{N}$$

   $$(D = \text{Major diameter of the thread})$$

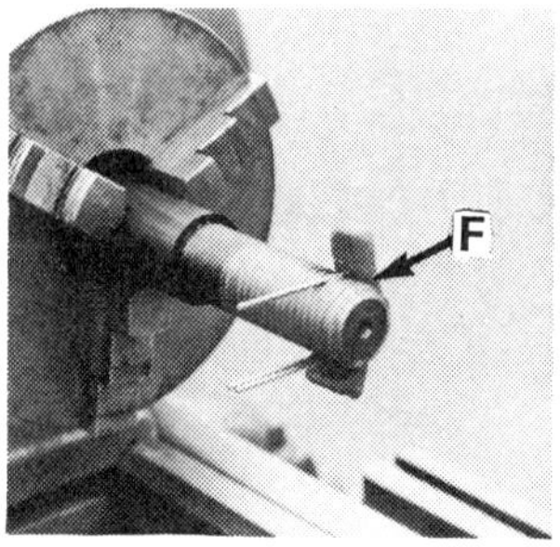

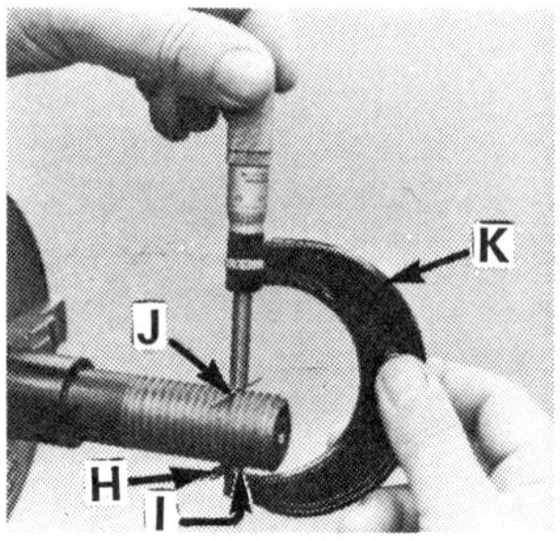

4. - **Place the three wires in the thread by:**

     (a) sticking the ends in a piece of plasticine (F)

     OR

     (b) holding two wires (H) in the bottom of the thread with the micrometer anvil (I).
   - **Put the single wire (J) in the top of the thread.**
   - **Measure over the wires with a micrometer (K).**
   - **Compare the reading with the calculated measurement over the wires.**

LAYOUT PROCEDURES

The purpose of laying out any work is to indicate the amount of material which is to be removed and to position holes, slots, etc. In order to be accurate, all layouts should be started from a finished edge or a "base" line.

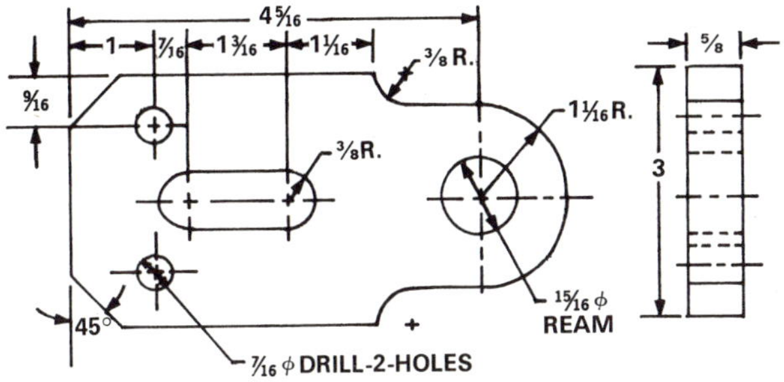

1. • **Although workpieces may vary considerably, the following sequence of layout operations can be followed for most jobs:**
   (1) Layout all lines parallel to an edge.
   (2) Layout all lines at 90° to the edge.
   (3) Layout all arcs and circles.
   (4) Layout the angular lines.

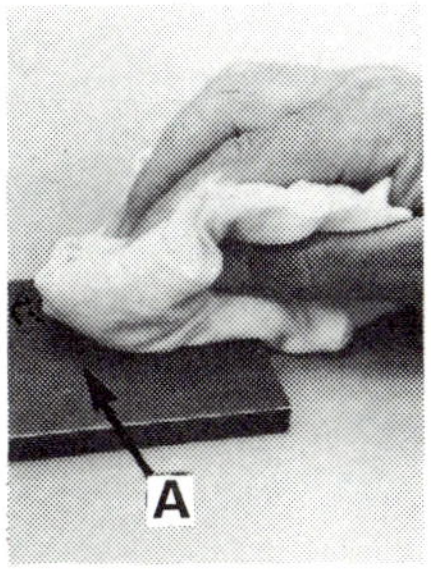

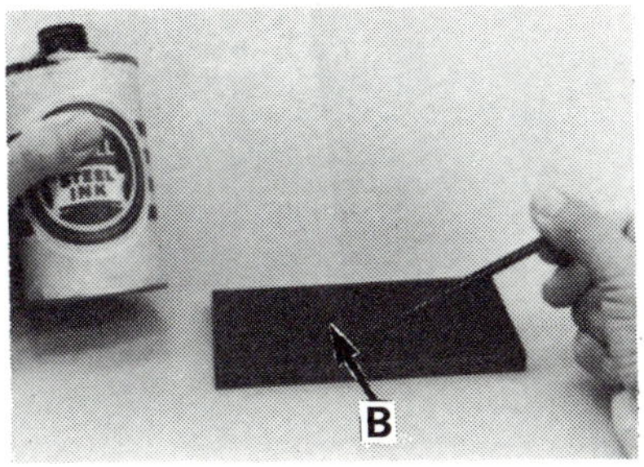

2. • **Remove all burrs from the work with a file.**
   • **Clean the surface of the work (A) thoroughly with a cloth to remove all dirt, grease, and oil. This will allow the layout dye to stick to the surface of the work.**
   • **Apply a** ***light coating*** **of layout solution (B) over the surface to be layed out.**

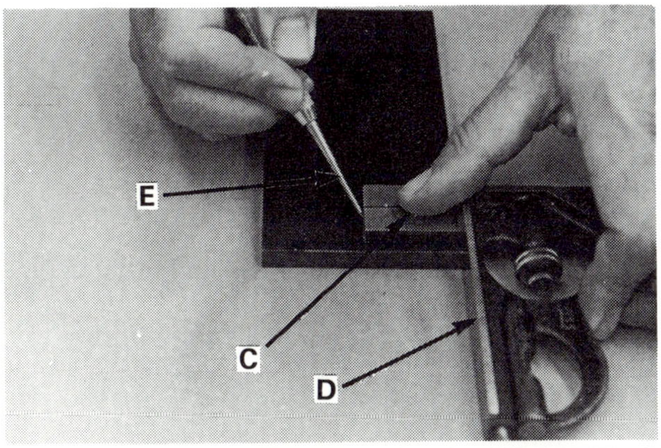

3. **COMBINATION SQUARE LAYOUT** (Center Lines)
   • **If possible, hold the workpiece in a vise to stop it from moving during the layout operation.**
   • **Extend the rule (C) beyond the body of the square (D) to approximately half the width of the work.**
   • **Hold the body of the square tightly against one edge with the thumb holding the blade flat on the work surface.**
   • **With a sharp scriber (E), held at an angle, draw a line along the end of the rule.**

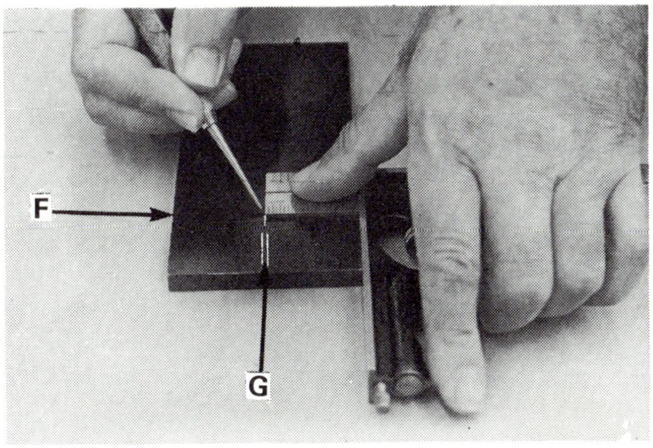

4. • **Place the square against the opposite edge of the workpiece (F).**
   • **Scribe a sharp line along the end of the rule beside the other line.**
   • **If both lines are not in the same place as shown in (G), adjust the distance the rule extends beyond the square until it is in the center of the two lines.**
   • **Rescribe lines from both sides of the workpiece.**

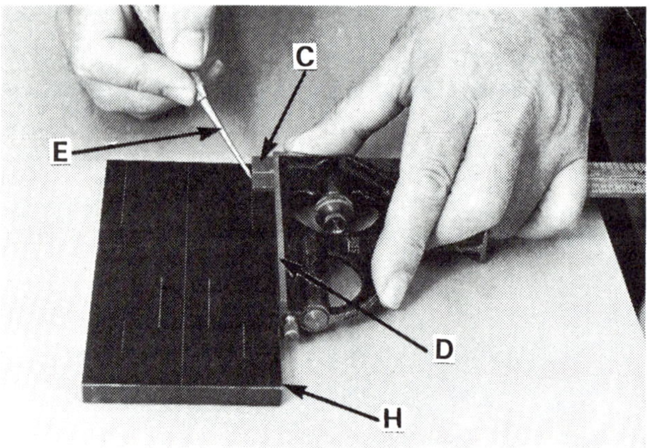

5. LAYING OUT HORIZONTAL LINES
- Extend the rule (C) the desired distance beyond the body of the square (D).
  NOTE: Be sure that only half the graduation line can be seen when locking the rule to the body.
- Hold the body of the square tightly against the machined edge (H).
- With a sharp scriber (E) held at an angle, draw a line along the end of the rule.
- Reset the rule and layout all lines parallel to the same edge.

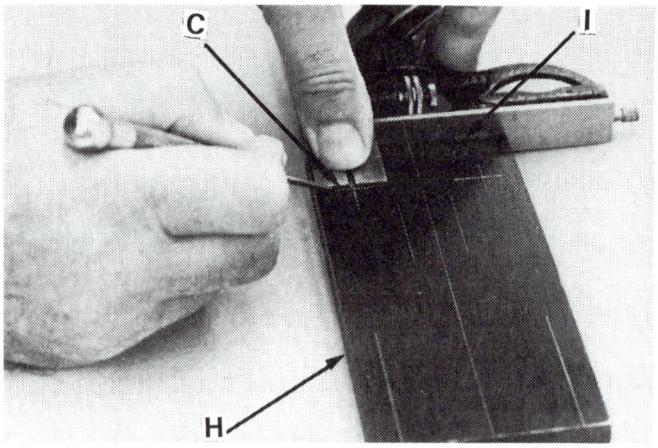

6. LAYING OUT VERTICAL LINES
- Extend the rule (C) the desired distance beyond the body of the square.
- Hold the body of the square tightly against edge (I). NOTE: Edge (I) must be at 90° to edge (H).
- With a sharp scriber held at an angle, draw a line along the end of the rule.
- Reset the rule and layout all lines parallel to edge (I).

7. SURFACE GAGE LAYOUT
- Clean the surface plate (J) with a cloth.
- Set a clean angle plate (K) on the surface plate.
- Place a 1" (25.4 mm) parallel (L) against the angle plate.
- Set the work on the parallel and against the angle plate so that end (M) is exactly even with the edge of the angle plate.
- Hold the work down against the parallel and fasten it against the angle plate with two clamps (N).

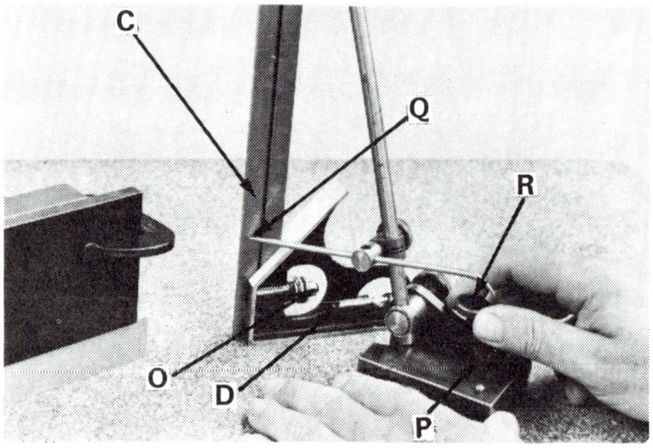

8.
- Set a combination square (D) on the surface plate.
- Loosen the lock nut (O) and be sure that the rule (C) is down against the surface plate.
- Tighten the lock nut.
- Set the surface gage (P) so that the scriber point (Q) is near the dimension required.
- Turn the adjusting screw (R) to make the final setting of the scriber.

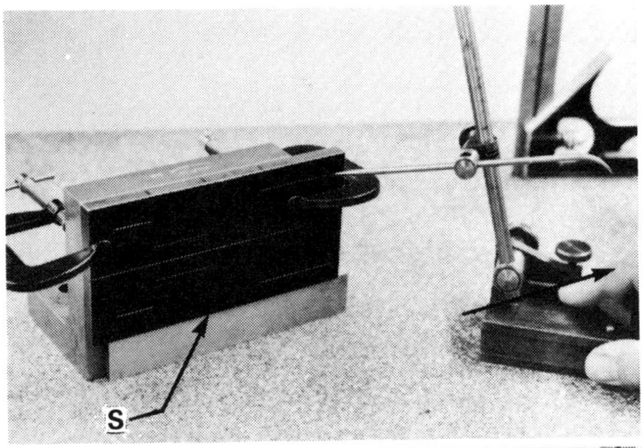

9. • Hold the surface gage down against the surface plate.
   • Draw the surface gage along the work in the direction of the arrows to scribe the lines. PUSHING will cause the scriber point to dig into the work and cause an inaccurate layout.
   • Reset the surface gage for each dimension.
   • Layout all lines which are parallel to edge (S).

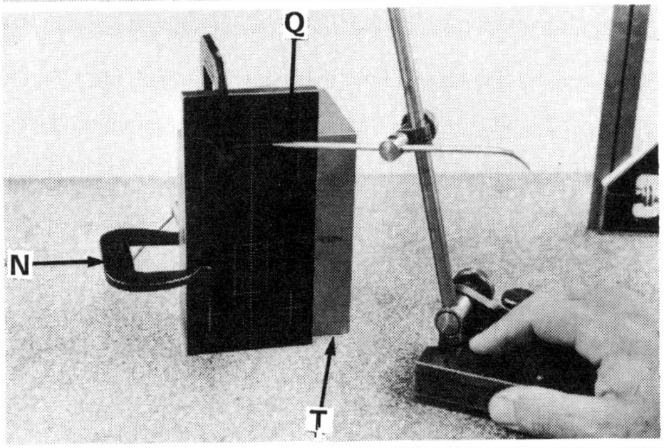

10. • Place a clamp (N) on the top of the work.
    • Remove the clamp from the left end of the work.
    • Set the angle plate on its left hand edge (T).
    • *For each dimension* set the surface gage scriber point (Q) with the rule on the square.
    • Layout all lines parallel to edge (T).

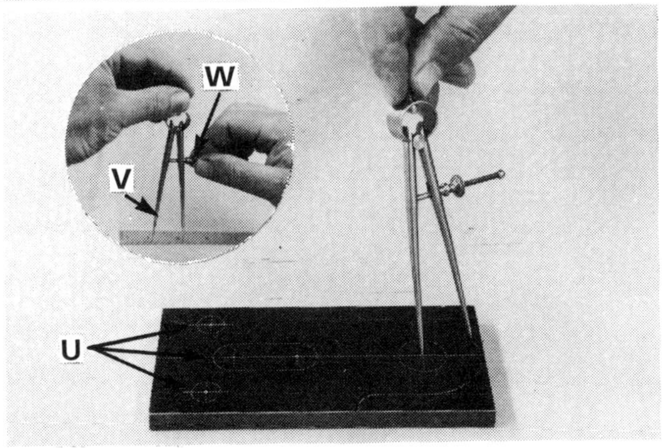

11. • Lightly prick punch the centers of all holes and arcs (U). (Use a magnifying glass to check the accuracy of the punch marks.)
    • Place one leg of the divider (V) in a main graduation on the rule.
    • Turn the adjusting nut (W) until the divider is set to the correct dimension.
    • Carefully layout all circles and arcs.

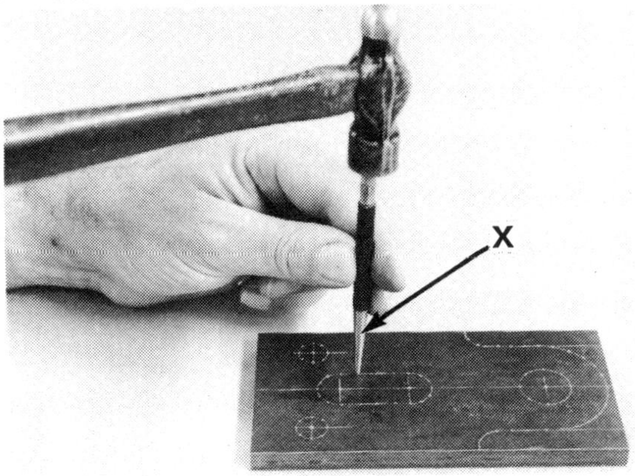

12. • To preserve the layout, use a sharp punch (X) and place a light punch mark about every ¼″ (6.35 mm) along each line.
    • Holes up to ¾″ (19.05 mm) diameter should use four punch marks around the circumference. Use eight punch marks for holes larger than ¾″ (19.05 mm) diameter.

The hand hacksaw is used to cut off work when it is inconvenient to use a power saw. Hacksaw blades are made of hardened high-speed molybdenum or tungsten-alloy steel.

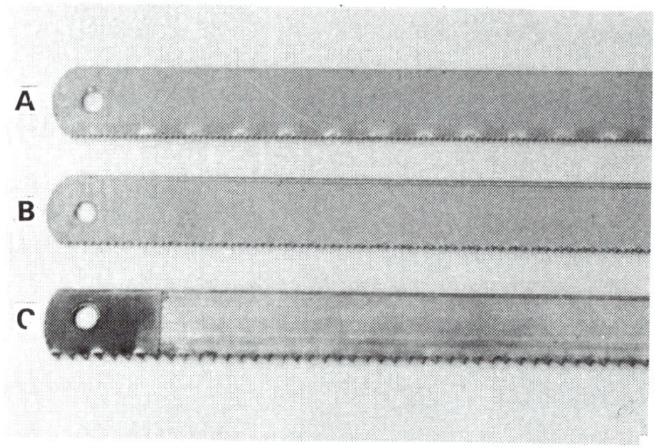

1. (A) **Fine pitch blades** (24 to 32 teeth per inch) are used for sawing brass, copper, and thin sections.
   (B) **Medium pitch blades** (18 teeth per inch) are used for tool steel and general sawing.
   (C) **Coarse pitch blades** (14 teeth per inch) are used when sawing large sections.

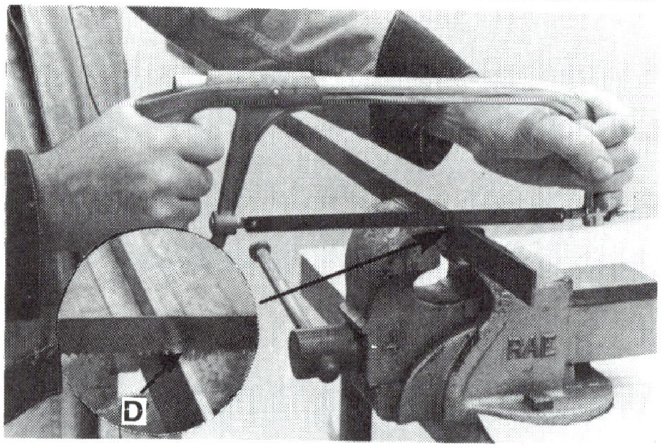

2. • Always have two saw teeth (D) in contact with the work.
   • Use as coarse a pitch blade as possible to provide good chip clearance and allow the saw to cut.
   • A fine pitch blade on a large section causes the teeth to clog and stops the saw from cutting.

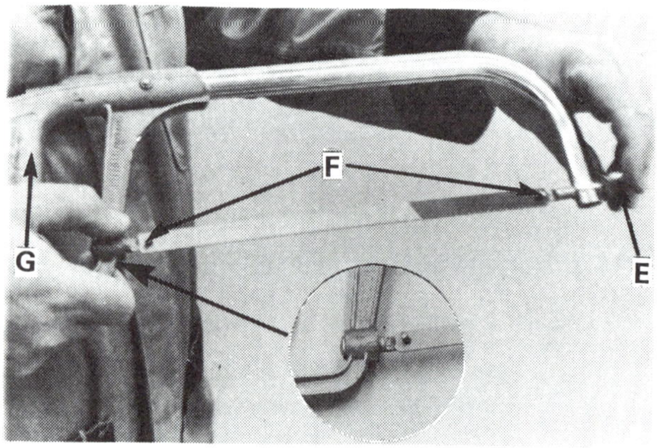

3. To insert a blade in the saw frame.
   (1) Loosen the wing nut (E).
   (2) Remove the old blade.
   (3) Insert the new blade onto the pins (F) being sure that the saw teeth point away from the handle (G).
   (4) Tighten the wing nut (E) as tight as possible with one hand.

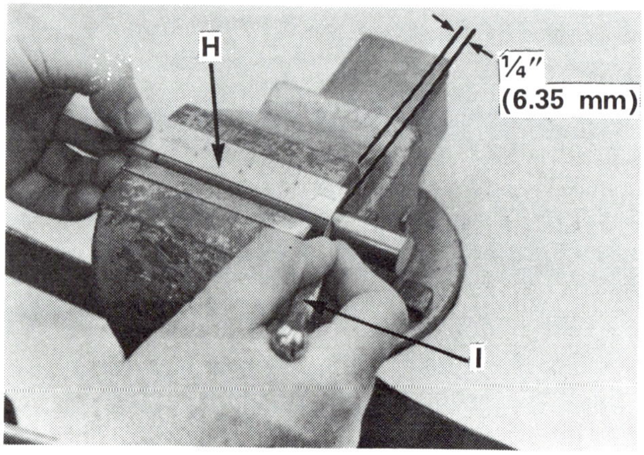

4. • Mark the length to be cut off with a rule (H) and scriber (I).
   • Use soft jaws in the vise if the work to be cut is finished.
   • Position the work so that the layout line extends about 1/4″ (6.35 mm) past the vise. (This will hold the work firmly and prevent it from springing.)

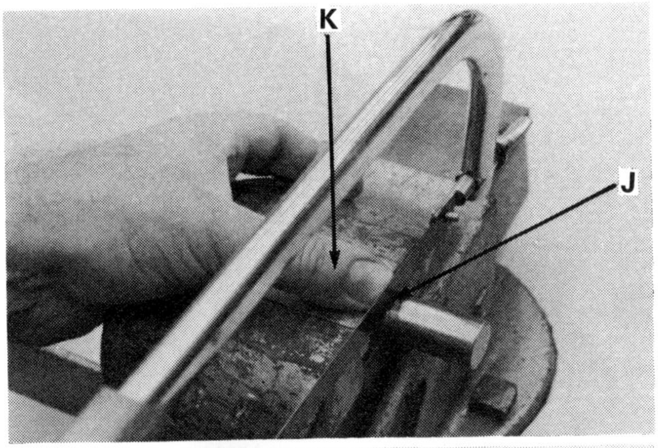

5. • Place the saw blade (J) on the scribed line. (A V-groove may be filed at the line for easier starting of the saw.)
   • Place a thumb (K) against the side of the saw blade to hold it in position.
   • Keep the thumb in position and begin sawing until the cut is about 1/16" (1.58 mm) deep.

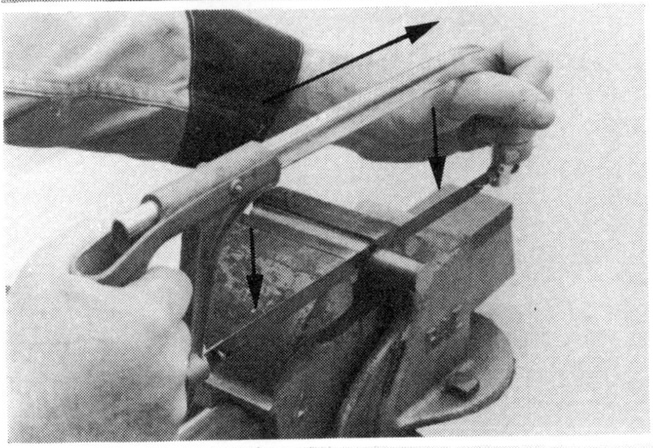

6. • Hold the saw firmly with two hands as shown.
   • Start sawing and apply down pressure on the forward stroke only.
   • Use about 50 strokes per minute when sawing.
   • Ease up on the sawing pressure when the piece is almost cut off.

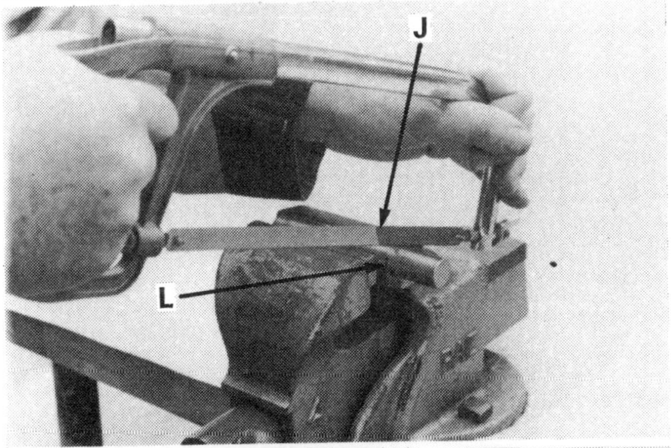

7. • If the saw blade breaks or becomes dull before the work is cut:
   (1) Replace the blade (J) with a new one. DO NOT start a new blade in an old cut. It will jam or break.
   (2) To overcome this, revolve the work ½ of a turn in the vise so that the old saw cut (L) is at the bottom.
   (3) Start the saw (as in Frame 5) and then cut the work (as in Frame 6).

8. SAWING THIN MATERIAL—less than the distance between two saw teeth.
   (1) Place the material (M) between two pieces of hard wood (N). (The wood on both sides of the thin material will prevent the saw from jamming and the work from bending.)
   (2) Grip the material and wood in the vise so that the position where the cut is to be made is close to the side of the vise.
   (3) Saw through the wood and thin material.

CUTTING INTERNAL THREADS

A tap is used to cut internal threads in a hole which has been drilled or bored to the tap drill size. Taps are available in a variety of type and sizes for hand or machine tapping.

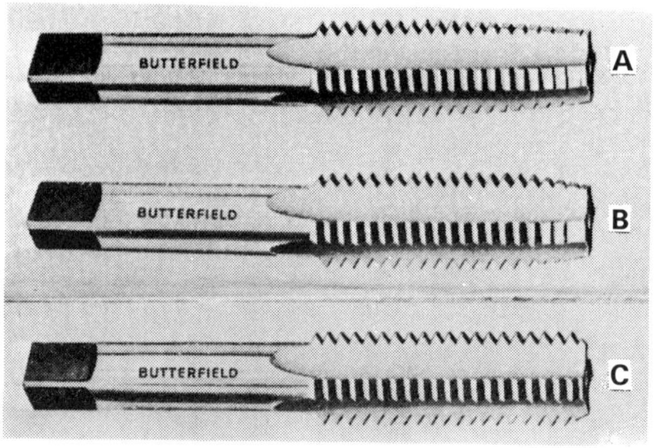

1. • Standard hand taps over ¼" diameter are available in sets of three.
   The *taper* tap (A) is used to start the thread in a hole.
   The *plug* tap (B) is used to follow the starting tap in a blind hole. It is often the only tap used for through holes.
   The *bottoming* tap (C) is used to finish the thread to the bottom of a blind hole.

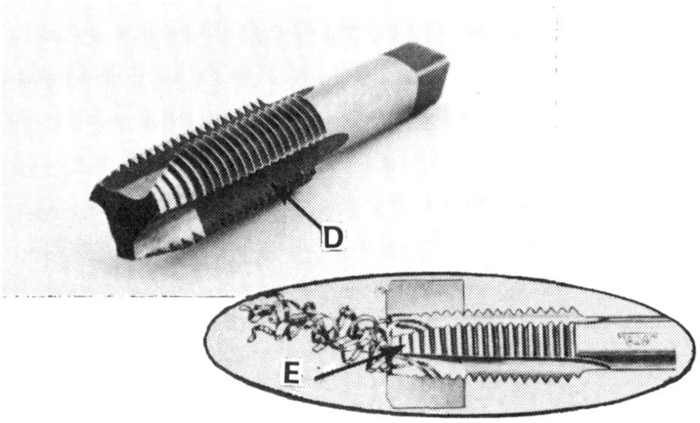

2. • Gun taps, used for hand tapping, generally have three flutes (D).
   • The chamfered end (E) has an angular flute which shoots the chips ahead of the tap. It can be used for through holes only and not for blind holes.

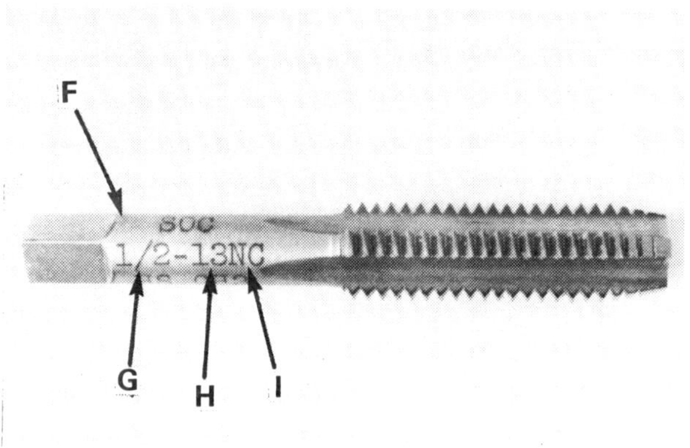

3. • The following information is generally stamped on the shank (F) of a tap:
   (1) Tap size or major diameter (G)—½".
   (2) Threads per inch (H)—13.
   (3) Type of thread (I)—U.N.C. (Unified National Coarse).
   • Left-hand taps have the marking L.H. in addition to the markings above.

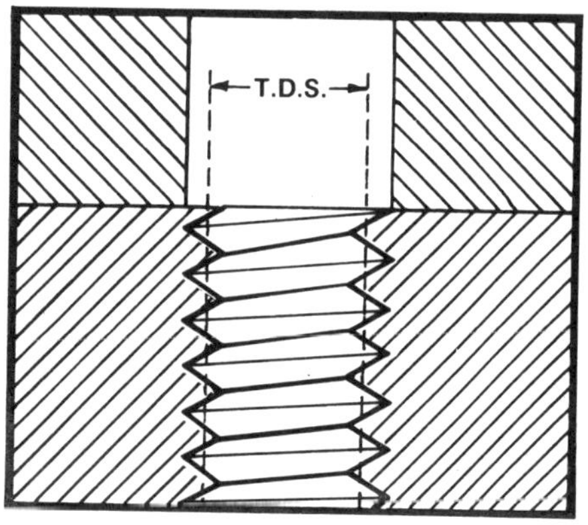

4. • Before a hole can be tapped, it must be drilled to the proper tap drill size.
   • Calculate the tap drill size for the tap to be used:

$$T.D.S. = D - \frac{1}{N}$$

T.D.S. = Tap drill size
D = Major diameter of the tap
N = Number of threads per inch

   • Drill the hole to the proper tap drill size.

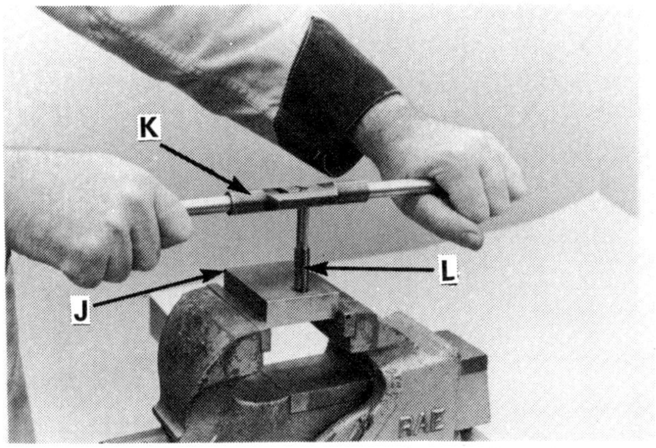

5. • Mount the work (J) in a vise.
   • Fasten the tap (L) in a tap wrench (K).
   • Place the tap in the hole and keep it as vertical as possible.
   • Apply cutting fluid to the tap.
   • Press downward on the tap wrench with equal pressure on both handles.
   • Turn the tap clockwise (for right-hand threads) about two turns; turn counterclockwise for left-hand threads.

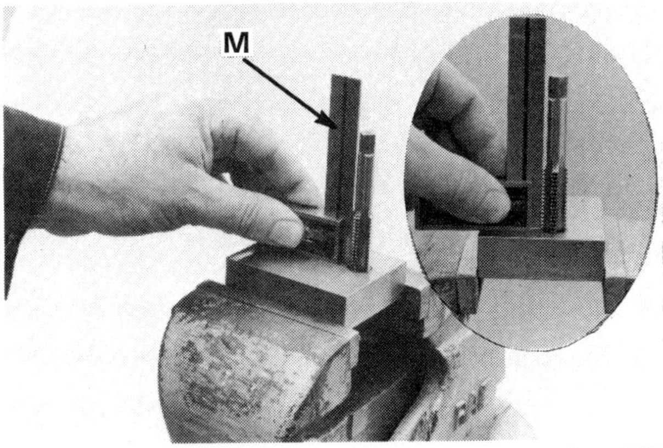

6. • Remove the tap wrench leaving the tap in the hole.
   • With a square (M), check the tap for squareness at two positions 90° to each other.
   • Note the alignment between the blade of the square and the shank of the tap.

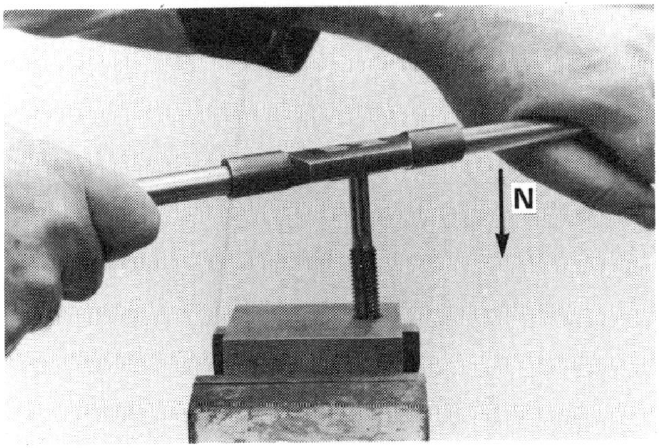

7. • If the tap is not square with the work, remove the tap from the hole.
   • Restart the tap and apply light pressure on the opposite side (N) to which the tap leans.
   • Turn the tap into the hole about three turns.
   • Recheck the tap for squareness and correct if necessary.
   • Tap the hole by turning the tap wrench alternately clockwise about ½ turn and then counterclockwise ¼ turn.
   • Use cutting fluid while tapping steel.

A

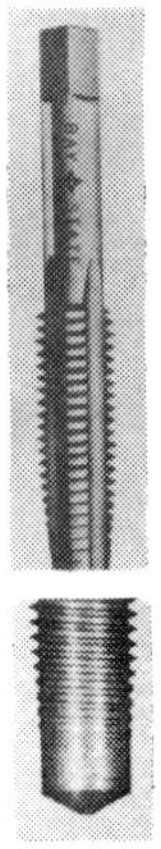

B

C

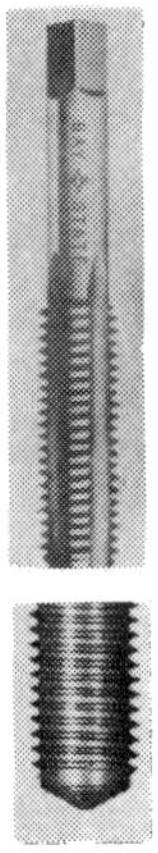

8. TAPPING BLIND HOLES
   • Measure the depth of the hole and mark it on the three taps with chalk or tape.
   • Start the taper tap (A) and thread the hole down to the mark.
   • Remove the tap and clean the chips from the hole.
   • Thread the hole with the plug tap (B).
   • Remove the tap and clean the chips from the hole.
   • Finish the thread with the bottoming tap (C).

CUTTING EXTERNAL THREADS

External threads may be cut on round work with a threading die mounted in a die stock. The die stock acts as a wrench to turn the die on the work. Dies use the same coding system as taps for identification.

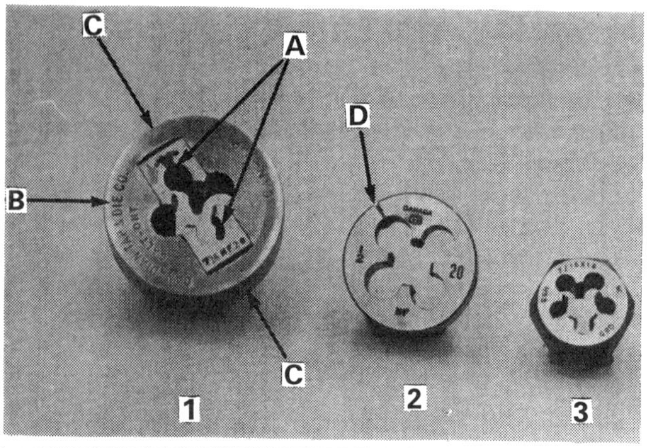

1. **TYPES OF DIES**
   - The *adjustable screw plate die* (1) consists of two die halves (A) held in a collet (B) by means of a threaded guide. Adjustment is made with two set screws (C).
   - The *adjustable split die* (2) is a more convenient form of die than the screw plate die. It is adjusted with one set screw (D).
   - The *solid die* (3) is used for chasing or recutting damaged threads. It is a finishing die and is not adjustable.

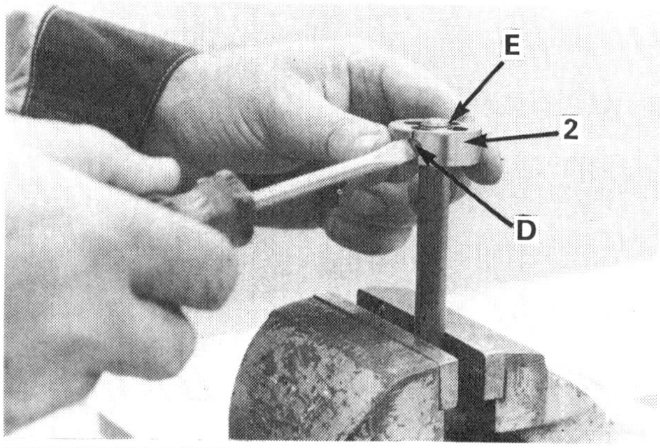

2.
   - To adjust the adjustable split die (2), turn the adjusting screw (D) *clockwise* to open the die.
   - Screw the die over a sample thread (E).
   - Turn the set screw (D) *counterclockwise* until the die fits snugly against the thread.
   - Remove the die from the sample thread.

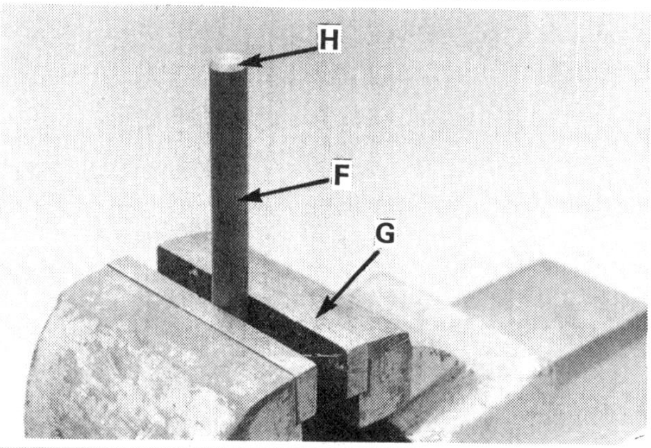

3. To cut an external thread on a round piece of work.
   - Place the workpiece (F) to be threaded in the vise (G). Be sure the end is chamfered (H) so that the die will start easily.
   - Lubricate the end of the work with cutting fluid.

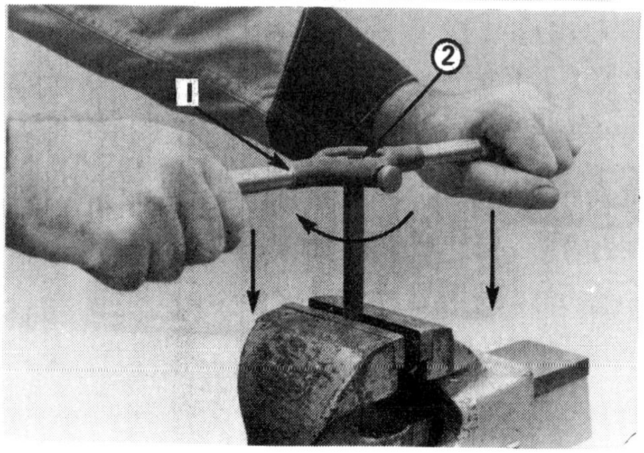

4.
   - Mount the split die (2) in a die stock (I).
   - Place it squarely on top of the workpiece with the tapered side of the die facing down.
   - Press down evenly on each handle and turn the die clockwise about two or three turns.
   - Check if the die has started squarely on the work. If it is not square, remove the die from the work.

5. • Start the die on the work.
   • Apply down pressure on the die stock to straighten the die on the work.
   • Turn the die clockwise for three or four turns while keeping the pressure on the desired side.
   • Check the die for squareness.
   • Repeat this procedure until the die is square.

6. • Apply cutting fluid.
   • Turn the die clockwise about one turn.
   • Turn the die counterclockwise about ¼ to ½ turn to break the chip.
   • Continue the procedure of turning the die clockwise one turn and then counterclockwise to break the chip.
   • Apply cutting fluid occasionally and cut the thread to the desired length.

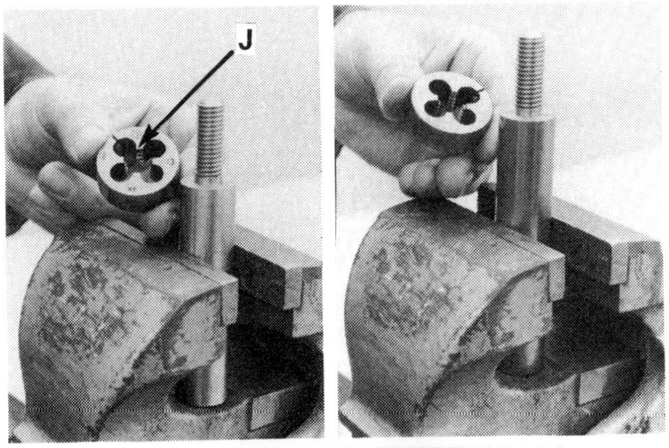

7. THREADING TO A SHOULDER
   • Start the die on the work with the tapered side (J) down.
   • Apply cutting fluid.
   • Cut the thread until the die is about 1/16" (1.58 mm) from the shoulder of the work.
   • Remove the die from the work.
   • Turn the die over so that the tapered side is up.

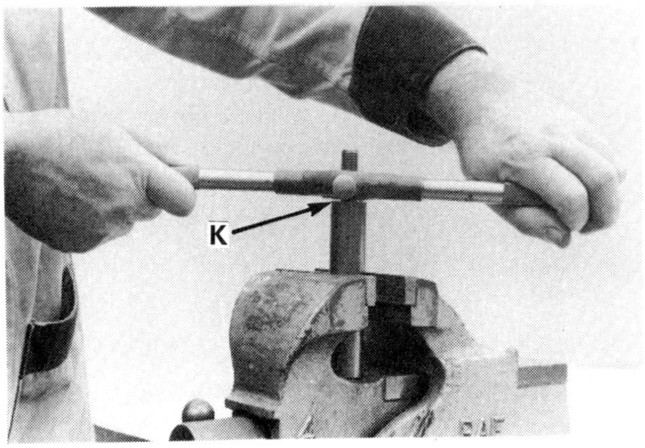

8. • Start the die onto the work and bring it near the bottom of the thread.
   • Cut the thread until the die almost touches the shoulder (K).
     CAUTION: If the die hits the shoulder, the work may be bent or the die damaged.
   • Remove the die and test the thread with a nut.

HAND REAMING

Reaming is performed when a hole must be finished accurately to size and shape and have a smooth finish. Although most reaming is done by machine, there are times when a hand reamer must be used to finish a hole.

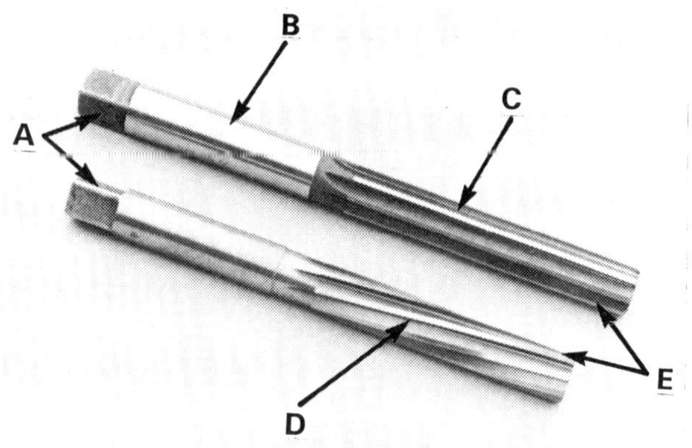

1. • Hand reamers are identified by the square (A) on the end of the shank (B).
   • Hand reamers with straight flutes (C) may be used for most hand reaming jobs.
   • Spiral flute reamers (D) are used in holes having keyways or other interruptions.
   • To permit the reamer to start easily, hand reamers are tapered slightly at the end (E) for a distance equal to the diameter.

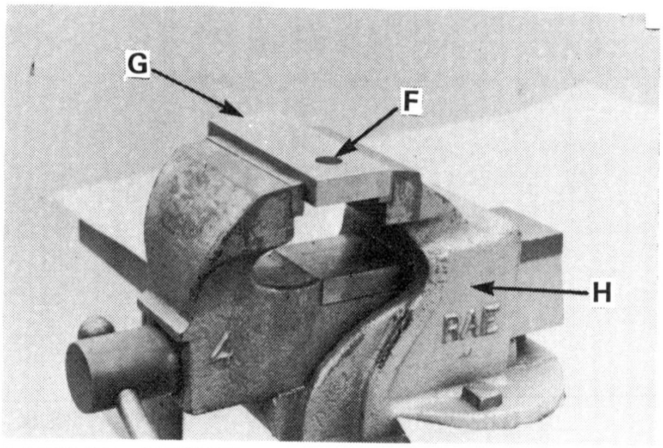

2. • The hole (F) in the workpiece (G) should be drilled and bored, or accurately transferred from another workpiece and then drilled.
   • For hand reaming, the hole (F) should be .005 to .010 smaller than the finished hole size.
   • Place the workpiece (G) in a vise (H).
   NOTE: Place soft jaws or protector strips over the vise jaws to protect finished work surfaces.

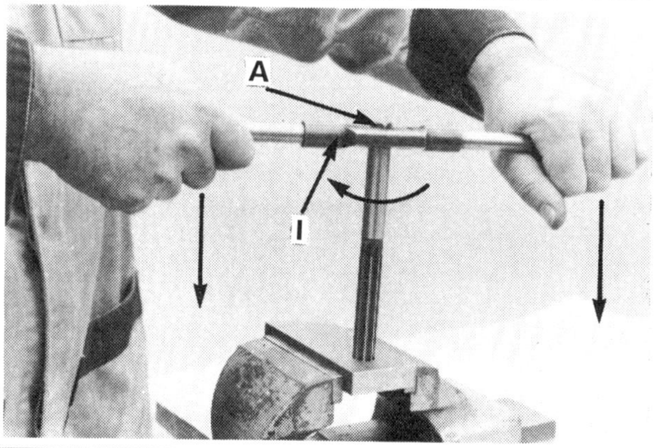

3. • Place the reamer in the hole.
   • Fasten a tap wrench (I) on the square end (A) of the reamer.
   • Apply cutting fluid to the reamer.
   • With equal down-pressure on both ends of the tap wrench, rotate the reamer clockwise until it starts cutting.
   • Check the reamer for squareness at several points.
   • Rotate the reamer slowly clockwise and apply light down-pressure to feed the reamer through the work.

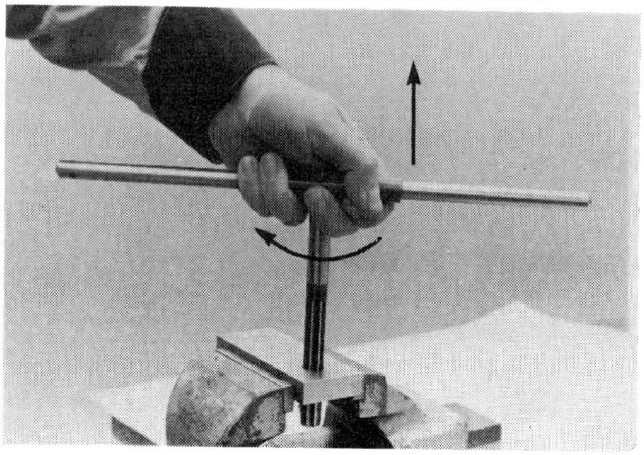

4. • Once the reamer is through the workpiece, grasp the tap wrench in the center with one hand.
   • Turn the reamer *clockwise* and lift up at the same time to remove the reamer from the hole.
   NOTE: Never revolve a reamer backwards or the reamer will be damaged.

DRILLING CENTER HOLES

Work that is to be machined between centers on a lathe must have a center hole drilled in each end to provide a bearing surface for the lathe centers. The center holes should be as smooth as possible to reduce the friction when the work revolves on the lathe centers.

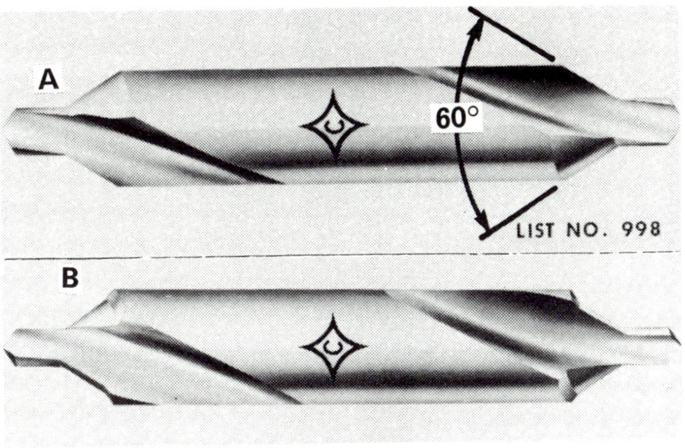

1. • The regular type center drill (A) has a 60° angular section with a small drill located on the end.
   • The bell type center drill (B) is similar to the regular type but it has a secondary bevel near the large diameter.

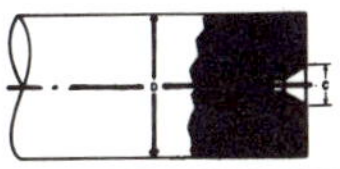

Size	Work Diameter D	Diameter of Countersink C	Drill Point Diameter	Body Size
1	3/16 to 5/16	3/32	3/64	1/8
2	3/8 to 1/2	9/64	5/64	3/16
3	5/8 to 3/4	3/16	7/64	1/4
4	1 to 1 1/2	15/64	1/8	5/16
5	2 to 3	21/64	3/16	7/16
6	3 to 4	3/8	7/32	1/2
7	4 to 5	15/32	1/4	5/8
8	6 and over	9/16	5/16	3/4

2. • Regular type center drill sizes range from #1 to #8.
   • Bell type center drills are prefixed by the number one (1) and range in sizes from 11 to 18.
   • *Body size* refers to the outside diameter of the center drill.

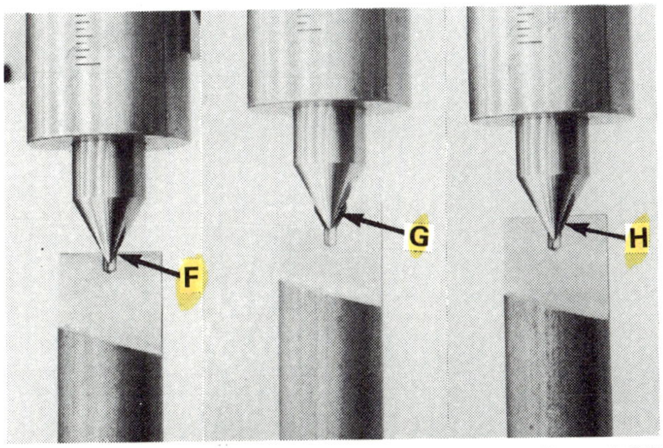

3. • The center hole (F) is too shallow and will not provide an adequate bearing surface.
   • The center hole (G) is too deep and will not allow the taper of the lathe center to contact the taper of the center hole.
   • The center hole (H), drilled to the proper depth, provides a good bearing surface for the lathe centers.

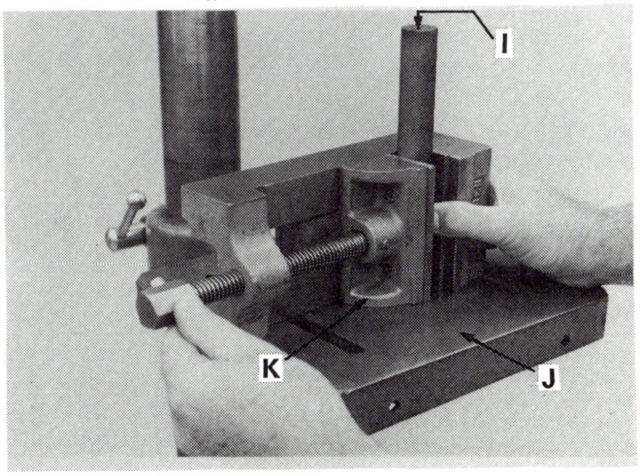

4. • Layout and center punch the center of each end of the workpiece (I).
   • Clean the drill press table (J).
   • Clean the vise and place it on its side (K).
   • Place the work in the vise in an upright position.
   • Press the work against the base of the vise and tighten the vise securely.

5. • Raise the table (J) so that the work is within about 2″ (50.8 mm) of the drill chuck.
   • Fasten a clamp (L) on the left side of the drill table to stop the vise from turning during the drilling operation.

6. • Select the proper size center drill (frame #2) to suit the diameter of the work.
   • Fasten the center drill (A) in the drill chuck (M) with no more than ¾″ (19.05 mm) extending beyond the chuck.
   • Calculate the drill press speed for the size of center drill being used.

$$\text{rpm} = \frac{4 \times CS}{\text{Drill Dia.}}$$

   • Change the belt (N) as indicated on the speed chart on the machine.

7. • Align the center punch mark directly below the point of the center drill.
   • Hold the vise firmly against the clamp.
   • Start the machine and gradually feed the center drill into the work.
   • Raise the drill from the work frequently and apply cutting fluid.

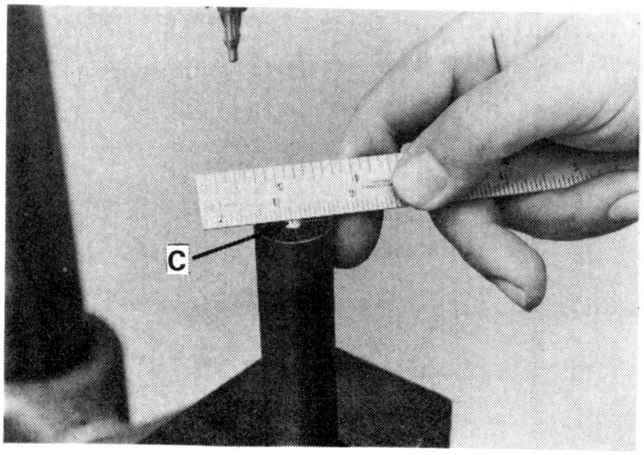

8. • Shut off the machine.
   • Measure the diameter at the top of the center hole (C) with a rule.
   • Continue drilling until the top of the countersunk hole is the correct size.
   • Inspect the hole for smoothness; if scratches or rings are present, apply cutting fluid and lightly bring the center drill into the hole.
   • Center drill both ends to the same diameter.

DRILLING A HOLE

When drilling holes, the work is generally held in a vise or placed on parallels and clamped to the drill press table. When drilling holes up to ½" (12.7 mm) diameter, the vise can be held by hand against a table stop. For holes over ½" (12.7 mm), clamp the vise or work to the drill press table.

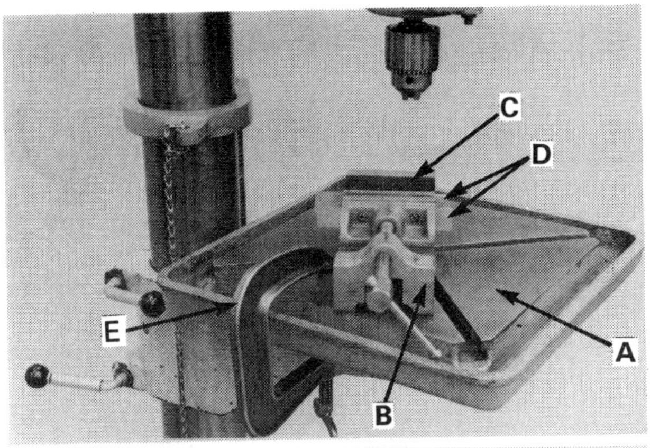

1.
- Clean the drill press table (A) and the vise (B).
- Place the work (C) on a set of parallels (D).
- Position the parallels so that the drill will clear them when it goes through the work.
- Tighten the vise securely.
- Tap the work down with a soft-faced hammer until the parallels are tight.
- Fasten a clamp (E) or stop on the left side of the table.

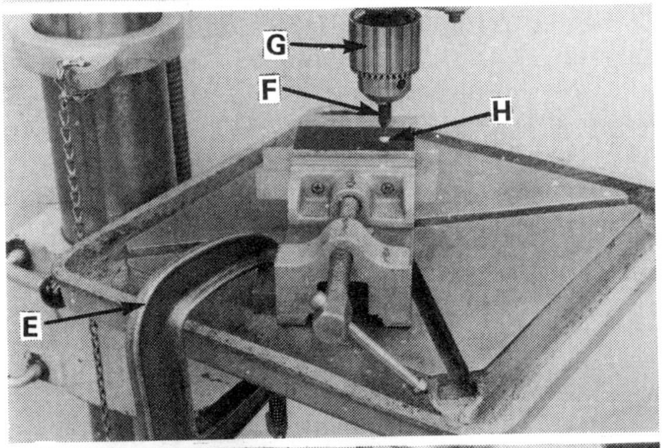

2.
- Fasten a center drill (F) in the drill chuck (G).
- Set the drill speed for the size of center drill.
- Place the vise against the clamp (E).
- Center the punch mark under the center drill.
- Start the drill press spindle.
- Drill the center hole (H) until one half of the tapered portion of the center drill enters the work.

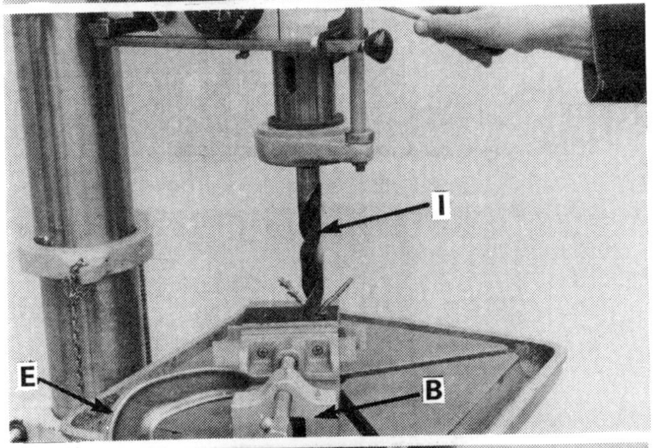

3.
- Mount the proper size drill (I) in the drill chuck or drill press spindle.
- Set the speed for the diameter of the drill.
- Bring the center hole under the drill point.
- Lower the drill into the center hole and lightly clamp the vise (B) to the table.
- Start the machine and feed the drill until it cuts into the workpiece about half the depth of the drill point.
- Tighten the clamp (E) securely.
- Apply cutting fluid and drill the hole.

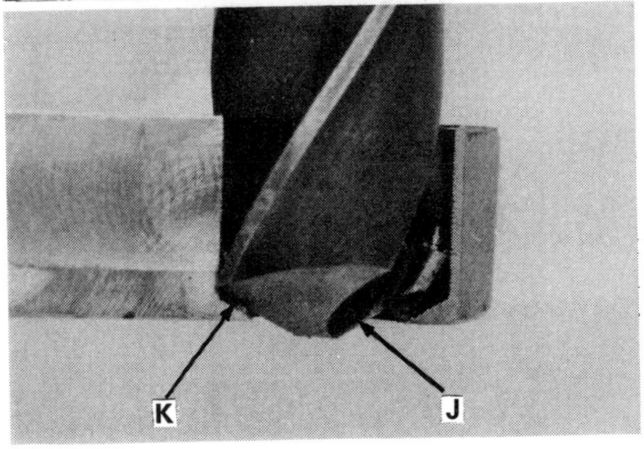

4.
- Ease up on the drilling pressure as the drill point (J) breaks through the work (K).

  This will prevent the drill from being pulled into the work and possibly breaking the drill or damaging the work.

DRILLING A LARGE HOLE

Before drilling a large hole, it is advisable to first drill a pilot hole slightly larger than the web of the drill which will be used to bring the hole to size. If the pilot hole is drilled too big, the large drill may chatter, drill the hole out-of-round, or spoil the top of the hole.

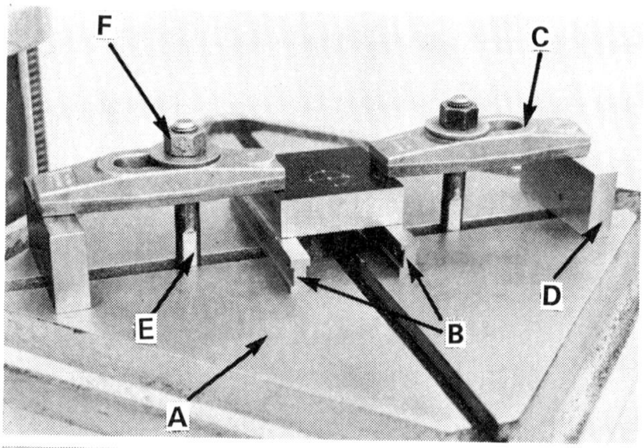

1. **WORK CORRECTLY CLAMPED**
   - **Clean the drill press table (A).**
   - **Place the work on a set of parallels (B) in the center of the table.**
   - **Place one end of each strap clamp (C) on a block (D) slightly higher than the workpiece.**
   - **Move both T-bolts (E) close to the work and lightly tighten the nuts (F).**
   - **The clamping pressure is applied to the work-piece and not the blocks.**

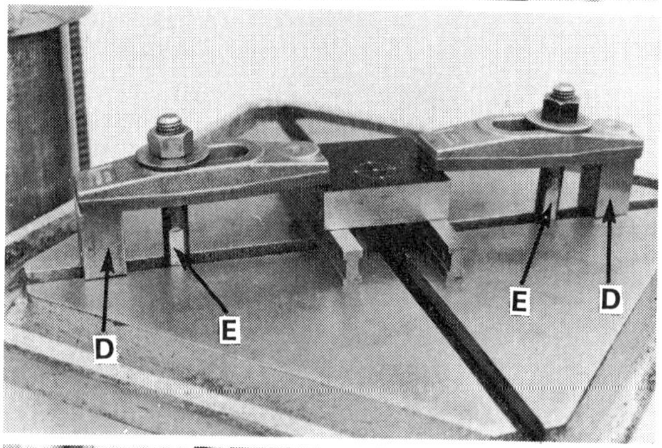

2. **WORK INCORRECTLY CLAMPED**
   - **The packing blocks (D) are too low.**
   - **The T-bolts (E) are too close to the blocks.**
   - **The clamping pressure is applied to the blocks and not to the workpiece.**

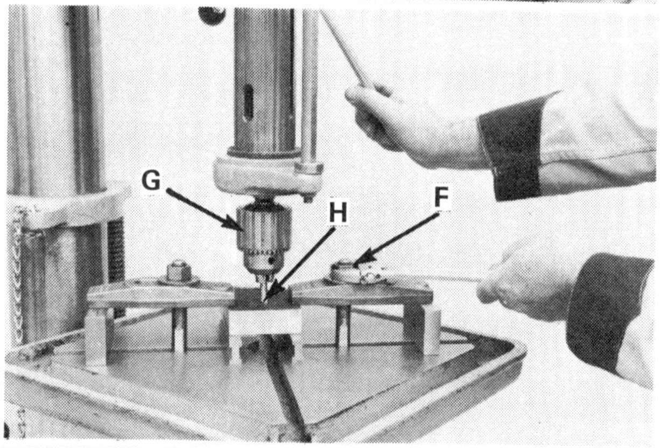

3. - **Adjust the table height so that the drill chuck (G) can be removed and the larger drill inserted in the spindle later without moving the table or the work.**
   - **Tighten the table lock.**
   - **Fasten a center drill (H) in the chuck.**
   - **Locate the center punch mark under the center drill point.**
   - **Spot the hole until the center drill enters to about one half of the angle section.**
   - **Tighten the nuts (F) securely.**

4. - **Remove the center drill from the chuck.**
   - **Measure the web thickness of the drill to be used.**
   - **Mount a pilot drill of this diameter (I) in the chuck.**
   - **Set the spindle speed and drill the pilot hole through the work.**
   - **Remove the drill from the chuck.**

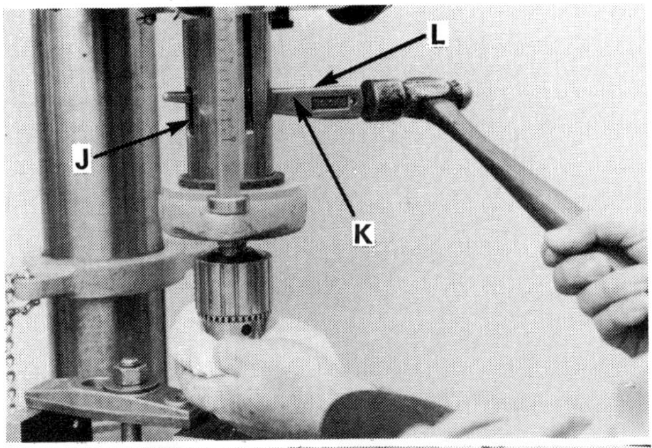

5. • Turn the drill press spindle by hand until the slot in the spindle aligns with the slot in the quill (J).
   • Select a proper size drill drift (K).
   • Insert the drift in the quill slot with its rounded edge (L) up.
   • Hold the chuck with one hand and tap the drill drift lightly with a hammer to remove the chuck from the spindle.

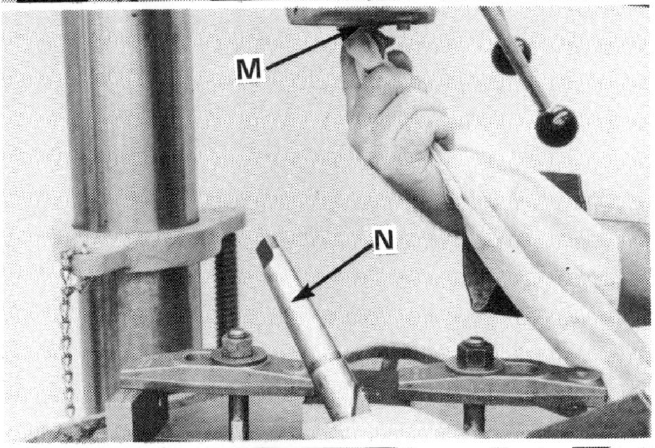

6. • DO NOT move the position of the table or workpiece.
   • Clean the drill press spindle taper (M).
   • Remove any burrs and clean the drill shank (N).

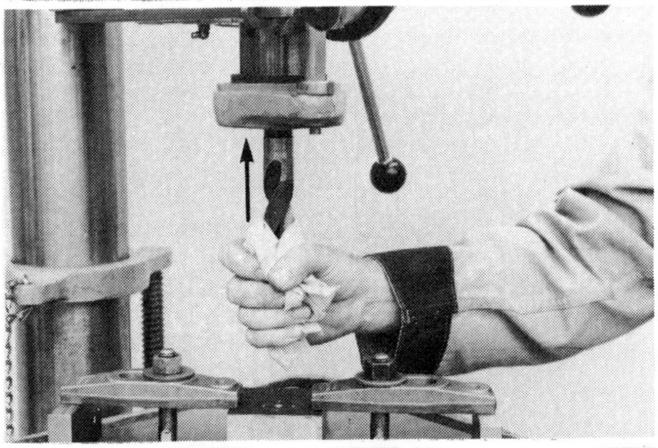

7. • Raise the drill spindle to the UP position.
   • Grasp the drill with a rag and fit the drill tang into the spindle slot.
   • Give the drill a quick snap upward to properly seat the tapered shank.

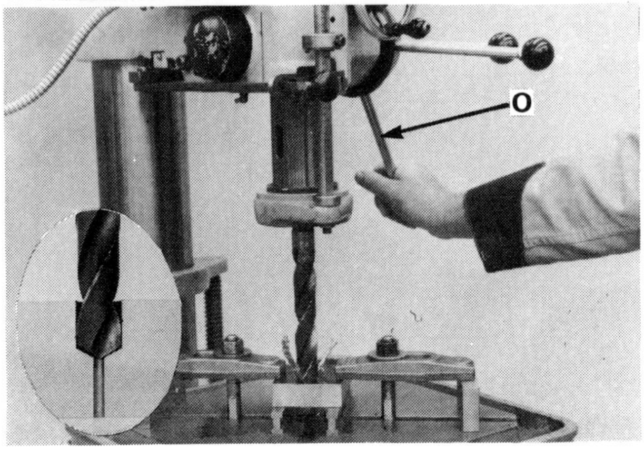

8. • Set the proper spindle speed.
   • With the downfeed handle (O) feed the drill into the workpiece carefully until it drills its full size.
   • Raise the drill occasionally and apply cutting fluid.
   • Release the drilling pressure slightly when the drill is breaking through the workpiece.

DRILLING TO A LAYOUT

Before work can be drilled, the position of the hole must be accurately laid out. The accuracy of the layout will generally determine the accuracy of the drilled hole.

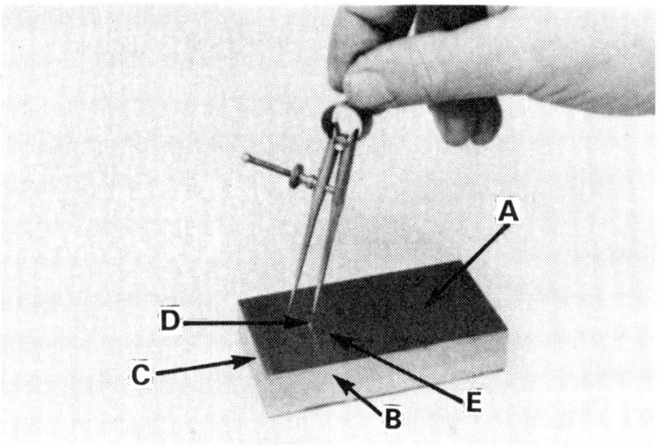

1. • **Clean the surface of the work (A) and coat it with layout dye.**
   • **Locate the center of the hole from the edges (B) and (C).**
   • **Lightly center punch where the two lines intersect (D).**
   • **Check the accuracy of the punch mark and correct if necessary.**
   • **With a divider, scribe a circle (E) the size of the hole required.**

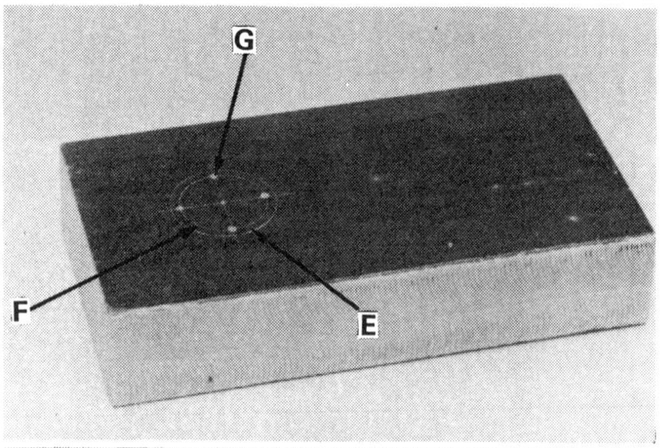

2. • **Set the divider and scribe a test circle (F) about 1/16" (1.58 mm) smaller than the first circle (E).**
   • **Place four light center punch marks (G) on the test circle where the lines intersect.**
   NOTE: On circles up to 3/4" (19.05 mm) diameter use four center punch marks and eight punch marks on larger circles.

3. • Clean the vise and set the work on parallels (H).
   • Tighten the vise and tap the work down with a soft-faced hammer until the parallels are tight.
   • Mount a center drill (I) in the drill chuck (J).
   • Set the correct drill press speed.
   • Spot the hole (K) so that only the point of the center drill enters the work.

4. • Mount the proper size drill in the spindle.
   • Set the proper drilling speed.
   • Feed the drill into the work until it cuts about one-half of its diameter (L).
   • Stop the machine.
   • Compare the impression left by the drill with the layout (E).

5. • If the drill is spotted to its full diameter, the drill point cannot be moved or corrected because:
   (a) the drill center point is guided by the edge of the hole (M).
   (b) the test circle is removed.
   (c) the center of the spotting can only be moved *before* the drill cuts its full size.

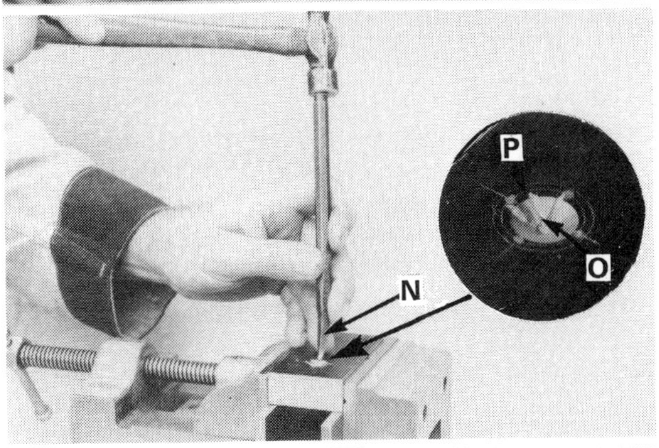

6. To move the center of the drilled spot
   - Hold a diamond or cape chisel (N) at the same angle as the spotted hole surface.
   - Cut three shallow grooves (O) on the heavy side (P) of the hole.

7. • Feed the drill into the work until the grooves are just removed.
   - Stop the drill and compare the spotting to the test circle (E).
   - If the spotting is not correct, cut three more light grooves on the heavy side of the hole.
   - Continue this procedure until the drill is drawn to the layout.

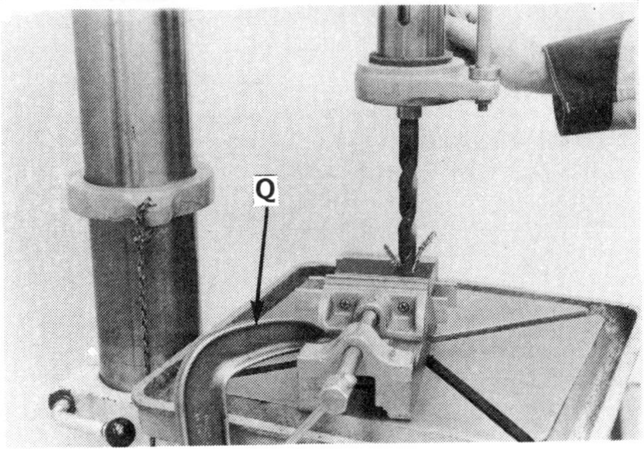

8. • Start the drill into the work. Then stop the machine.
   - Keep the drill in the work and fasten the vise to the table securely with a clamp (Q).
   - Apply cutting fluid and drill the hole through the work.
   - Release the feed pressure when the drill is breaking through the workpiece to prevent the drill from catching.

DRILLING ROUND WORK

In order to drill a hole through the center of round work, the center punch mark must be set up so that the drill will pass through the exact center of the work. The work is usually placed in a V-block which is held in a vise or clamped to the drill press table.

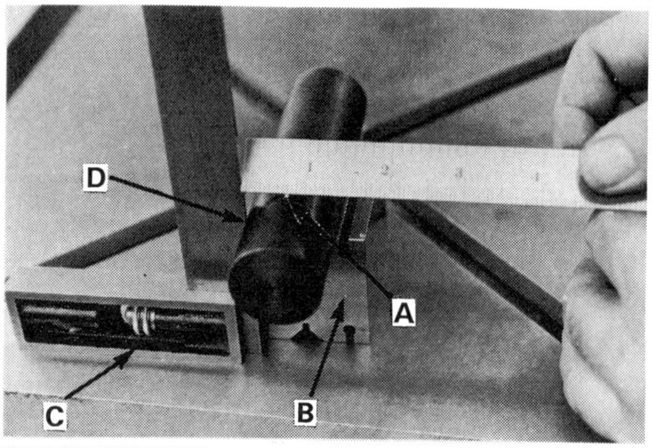

1. • Layout the position of the hole (A).
   • Set the work in a V-block (B).
   • Place a square (C) on the drill press table and bring the blade into contact with the work (D).
   • Measure the distance from the edge of the blade to the center punch mark (A).
   • Turn the workpiece in the V-block until the distance from the blade to the center mark (A) is half of the diameter of the work.

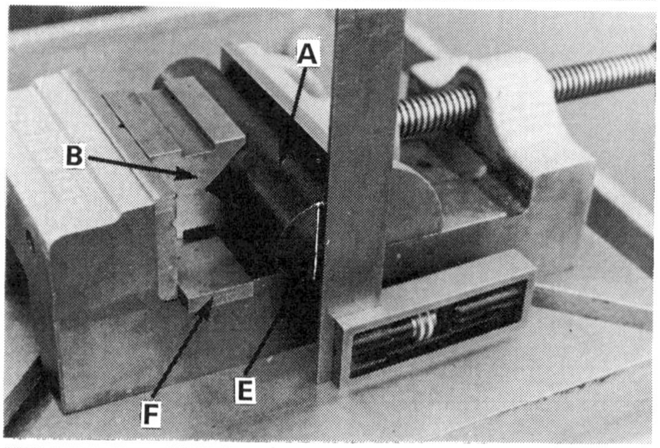

2. • Scribe a line (E) on the end of the work in line with the center punch mark (A).
   • Set a V-block (B) on a parallel (F) in a vise and tighten the vise slightly.
   • Place a square on the table and bring the blade close to the layout line (E).
   • Revolve the work in the V-block until the layout line is parallel with the blade of the square.
   • Tighten the vise securely.

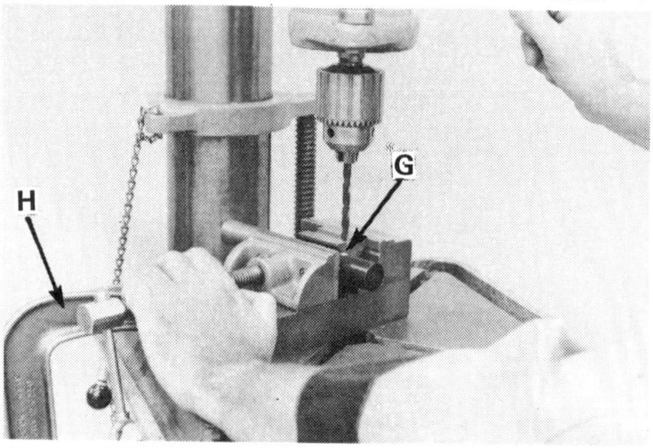

3. • With a center drill spot the hole (G).
   • Mount the proper sized drill in the drill chuck.
   • Set the correct spindle speed for drilling.
   • Clamp a table stop (H) on the left side of the table.
   • Align the drill to the center of the hole.
   • Apply cutting fluid and drill the hole.
   • Release the feed pressure just as the drill is breaking through the work.

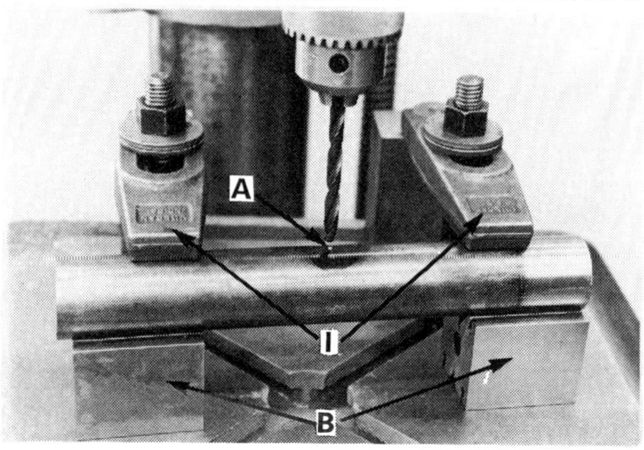

4. DRILLING LARGE DIAMETER WORK
   • Layout and center punch the hole location (A).
   • Set the work on a matched set of V-blocks (B).
   • Align the center punch mark (A) using the method in Frame #1 and 2.
   • Place a center drill in the drill chuck and align the punch mark with the drill point.
   • Tighten the clamps (I) on the work over the V-blocks.
   • Spot the hole with the center drill.
   • Drill the hole with the correct size drill using cutting fluid and proper drilling technique.

Three common methods are used to transfer hole locations from one part to another part. These are:

1. The use of one part as a master plate and the transferring of the holes by spotting with a drill.
2. The use of transfer punches.
3. The use of transfer screws.

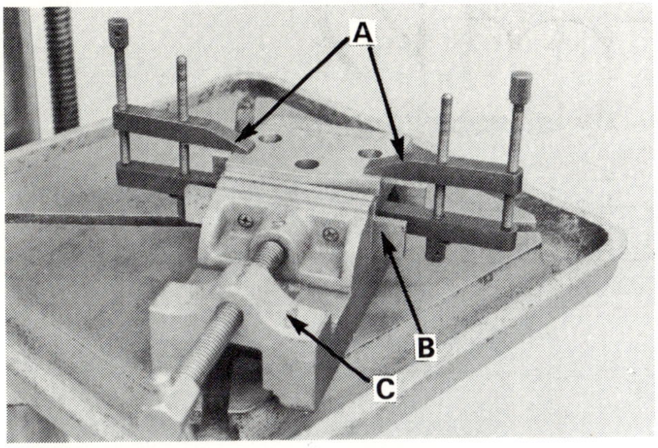

1. SPOTTING WITH A DRILL
   - Remove all burrs from both parts.
   - Place the part with the holes on top of the other part.
   - Accurately align the edges of both pieces.
   - Tighten the clamps (A) securely.
   - Clean the vise and set the work on parallels (B).
   - Tighten the vise (C) and tap the work down until the parallels are tight.

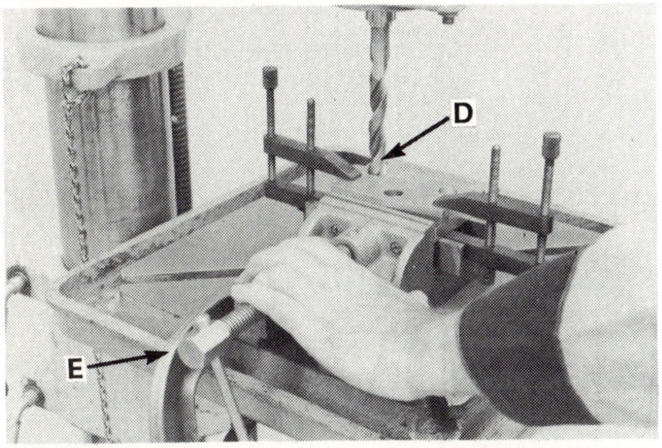

2.
   - Select a drill (D) which is the size of the hole to be transferred.
   - Mount the drill in the chuck or drill press spindle.
   - Position the vise so that the drill fits into one of the holes to be transferred.
   - Fasten a clamp (E) on the left side of the table.
   - Spot each hole until the full point of the drill has entered the work.

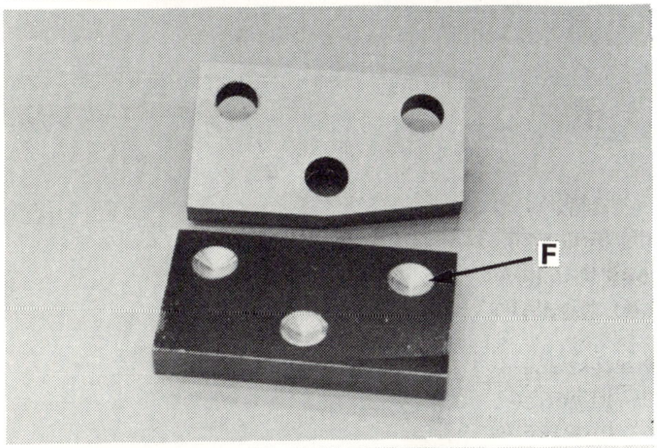

3.
   - Remove the clamps.
   - Remove the top or master plate.
   - Fasten the plate with the spotted holes (F) in the vise.
   - Drill the spotted holes to the required diameter and depth.

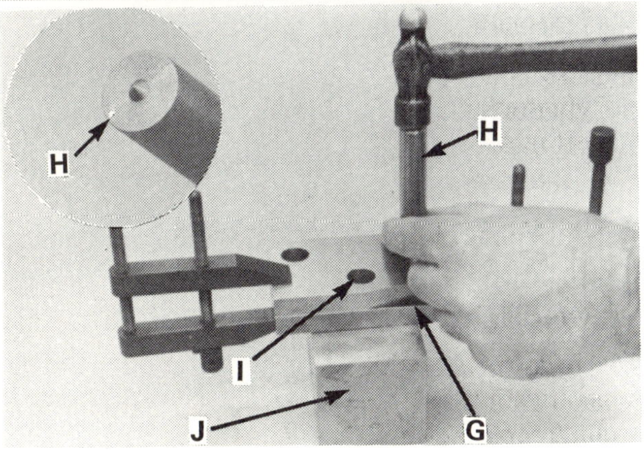

4. USING TRANSFER PUNCHES
   - Remove the burrs and clean both parts.
   - Apply layout dye to the surface of part (G).
   - Align both parts accurately and clamp them together securely.
   - Select a transfer punch (H) which is the same size as the holes (I) to be transferred.
   - Place the work assembly on a metal block (J).
   - Hold the transfer punch vertically in each hole and strike it lightly with a hammer.

5. • Remove the clamps from the work.
   • Separate both parts.
   • Use a sharp prick punch to deepen the transfer marks (K) slightly.

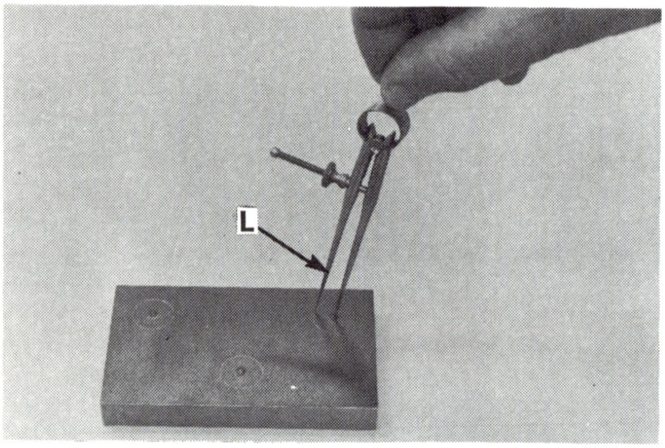

6. • With a divider (L) draw a test circle $\frac{1}{16}''$ (1.58 mm) smaller than the required diameter.
   • Place four light prick punch marks around the test circle.
   • Reset the dividers and draw a circle the diameter of the hole.
   • Drill the holes to size as outlined in Drill Press Operation #4.

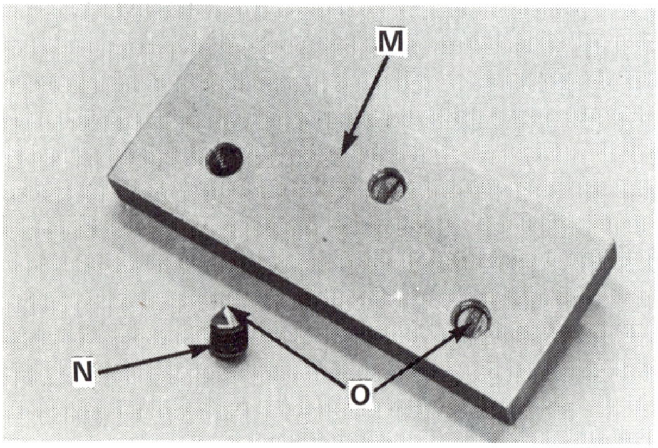

7. TRANSFERRING THREADED HOLES
   • Remove all burrs and clean the surface (M).
   • Select transfer screws (N) to fit the threaded holes.
   • Thread the transfer screws into the holes.
   • Have the transfer screw center point (O) extending about $\frac{1}{32}''$ (.79 mm) above the surface (M).

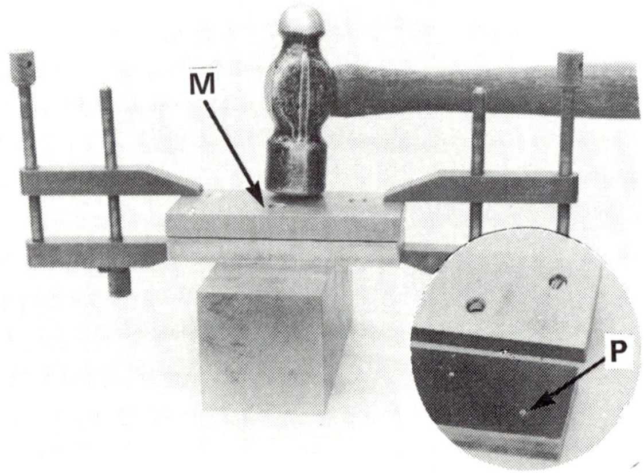

8. • Align the edges of the two pieces of work and clamp them together.
   • Place them on a solid block and sharply strike the part (M) with a hammer.
   • Remove the clamps.
   • Use a sharp prick punch to deepen the transfer screw marks (P) slightly.
   • Scribe test circles for the holes to be drilled.
   • Drill the holes to size as outlined in Drill Press Operation #4.

After a hole has been drilled, it may be finished round, smooth and to an accurate size by reaming. Reaming may be performed on a drill press as either a machine or hand operation.

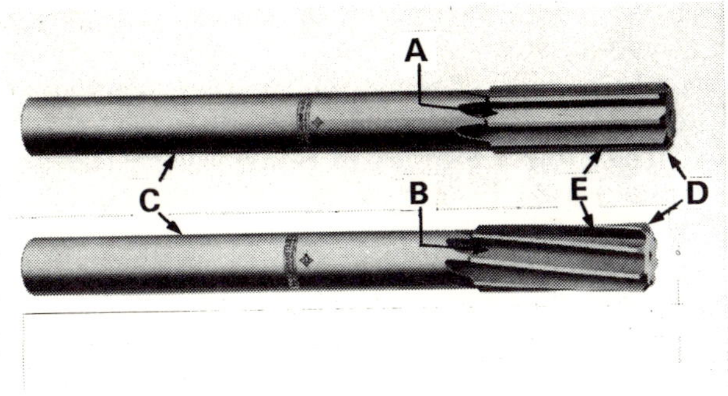

1. • Machine or chucking reamers with straight flutes (A) are used for general work.
   • Reamers with helical flutes (B) are used in holes having keyways or grooves.
   • The straight shanks (C)are held in a drill chuck.
   • The 45° chamfer (D) creates the cutting edge.
   • The margin (E) does not cut and only steadies the reamer in the hole.

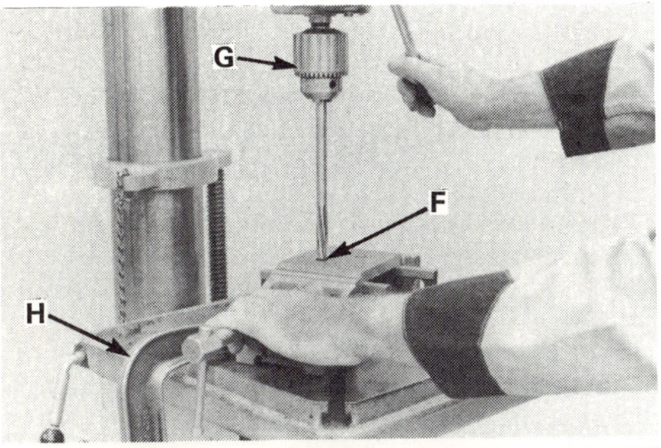

2. REAMING HOLES UP TO ½" (12.7 mm) DIAMETER
   • Set the work in the vise.
   • Drill a hole (F) 1/64" (.39 mm) under the finished size required.
   • Mount a machine reamer in the drill chuck (G).
   • Set the drill press speed to ¼ of the drilling speed.
   • Align the drilled hole with the reamer.
   • Hold the vise against the table stop (H).
   • Apply cutting fluid and ream the hole.

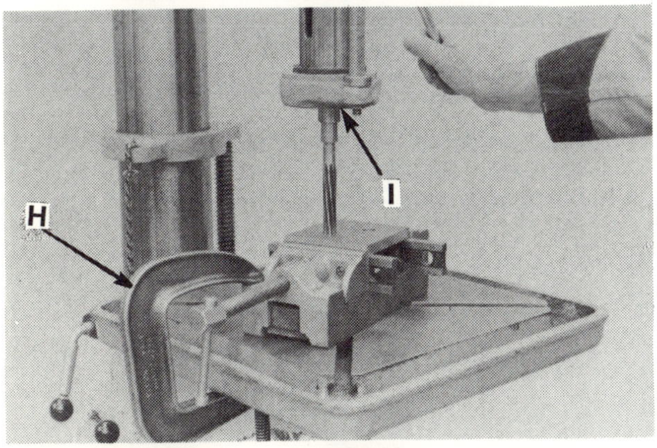

3. REAMING HOLES OVER ½" (12.7 mm) DIAMETER
   • Drill a hole 1/32" (.79 mm) under the finished size of the hole.
   • Fasten the vise to the table with a clamp (H) while the drill is in the hole.
   • Mount a machine reamer in the spindle (I) without moving the drill press table or work.
   • Set the drill press speed for reaming.
   • Apply cutting fluid and feed the reamer into the hole at a steady rate.

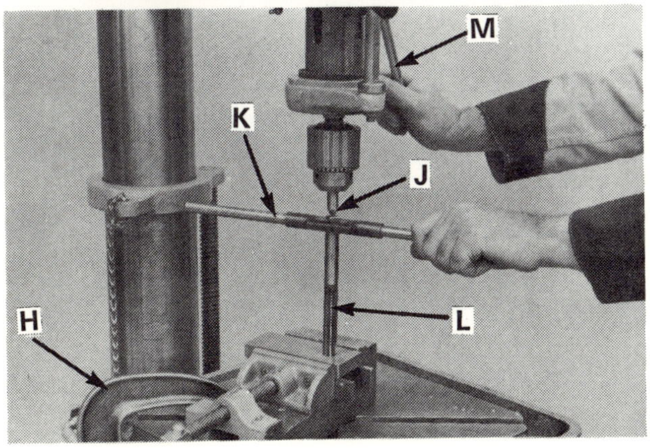

4. HAND REAMING
   • Drill a hole .005-.010 (.12-.25 mm) undersize.
   • Fasten the vise to the table with a clamp (H) while the drill is in the hole.
   • Mount a stub center (J) in the drill chuck.
   • Fasten a tap wrench (K) on the hand reamer (L).
   • Place the reamer in the drilled hole.
   • Engage the stub center in the reamer center hole.
   • With the downfeed lever (M) apply slight pressure while turning the reamer by hand.
   • Apply cutting fluid and ream the hole.

TAPPING A HOLE

Internal threads can be produced on a drill press by hand or with the use of a tapping attachment. Machine taps should be used when tapping with power equipment.

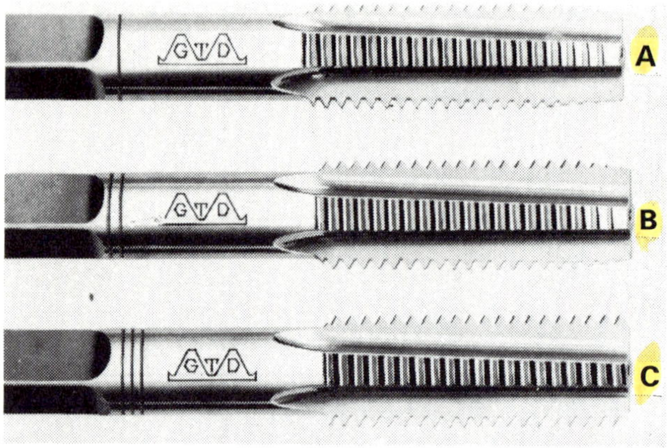

1.
   - Standard taps over ¼″ diameter are available in sets of three.
   - The taper tap (A) is used to start the thread in the hole.
   - The plug tap (B) is used to follow the starting tap in a blind hole. It is often the only tap used for through holes.
   - The bottoming tap (C) is used to finish the thread to the bottom of a blind hole.

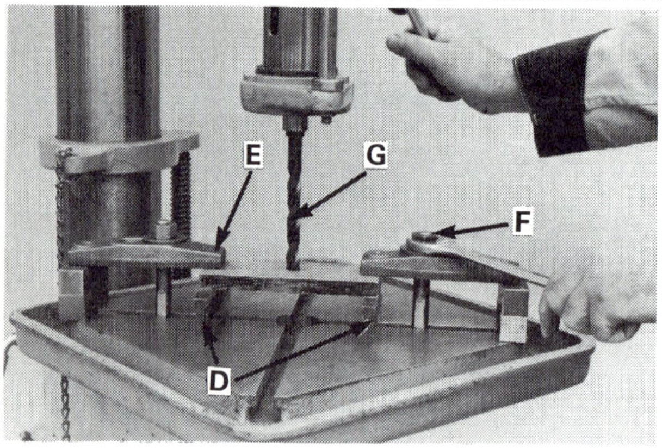

2.
   - Hold the work on parallels (D) with suitable clamps (E).
   - Adjust the table height so that the tap drill may be inserted later without moving the table or work.
   - Lock the table.
   - Center drill the hole and lightly tighten the clamping nuts (F).
   - Mount the tap drill (G) in the machine without moving the table or work.
   - Feed the revolving drill into the work and tighten the clamping nuts securely.
   - Apply cutting fluid and drill the hole.

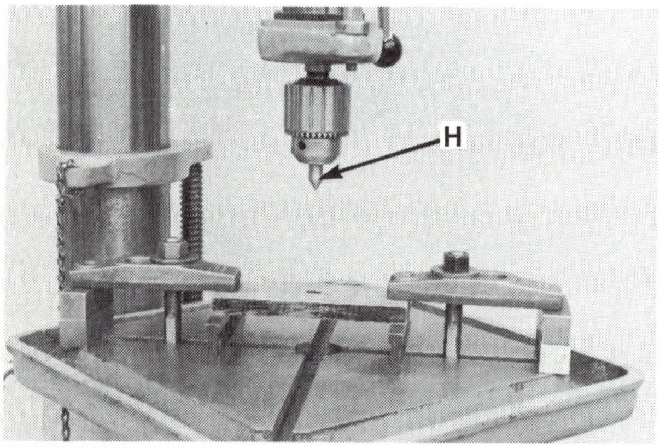

3. TAPPING BY HAND
   - Remove the tap drill without moving the table or work position.
   - Mount a stub center (H) in a drill chuck.
     The stub center should extend about 1″ (25.4 mm) beyond the drill chuck.

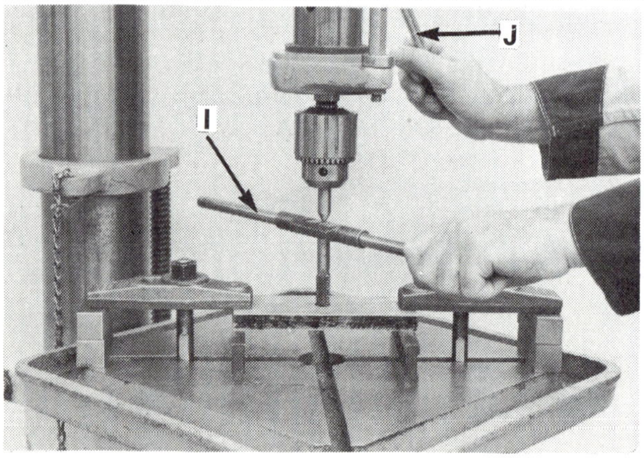

4.
   - Fasten a tap wrench (I) on the tap.
   - Place the tap in the hole and support its top with the stub center.
   - Apply down pressure with the downfeed handle (J) and turn the tap clockwise two or three turns.
   - Apply cutting fluid to the tap.
   - Maintain light pressure on the downfeed handle and thread the hole.
   - Reverse the tap after each revolution to break the chip.

5. TAPPING UNDER POWER
- Remove the drill chuck and clean the hole in the drill press spindle (K).
- Mount the tapping attachment (L) in the spindle.
- Position the stop rod (M) against the column (N).
- Set the proper speed for tapping (about $\frac{1}{4}$ of the drilling speed).

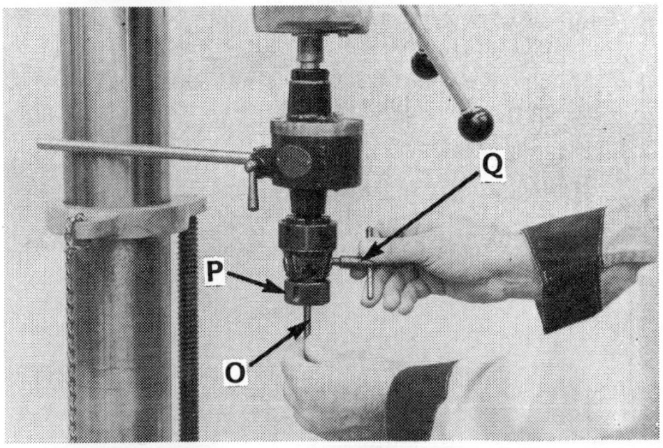

6.
- Insert the shank of the tap (O) into the chuck of the tapping attachment.
- Tighten the lock nut (P) by hand.
- Turn the wrench (Q) to tighten the jaws on the square of the tap.
- Tighten the lock nut securely with the proper wrench.
- Start the drill press and check the tap for trueness.

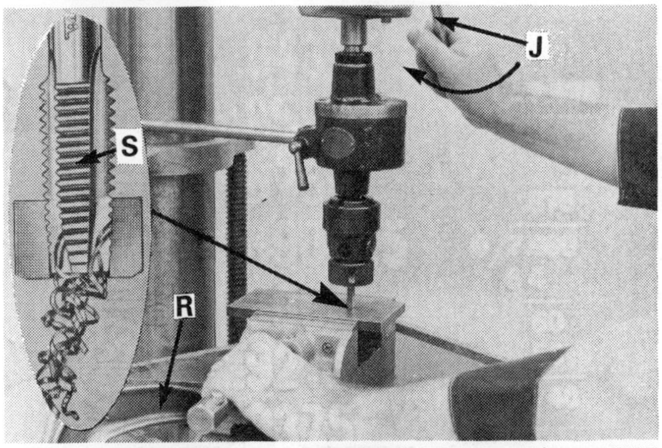

7.
- Align the hole to be threaded with the tap.
- For taps up to $\frac{5}{16}$" (8 mm) diameter, the vise may be held by hand against a stop (R).
- For taps over $\frac{5}{16}$" (8 mm) diameter, the vise must be clamped to the table.
- Start the drill press.
- Apply cutting fluid to the tap.
- Feed the tap through the hole with steady pressure on the downfeed handle (J).
- A gun tap (S) is designed to shoot the chips out of the hole ahead of the tap.

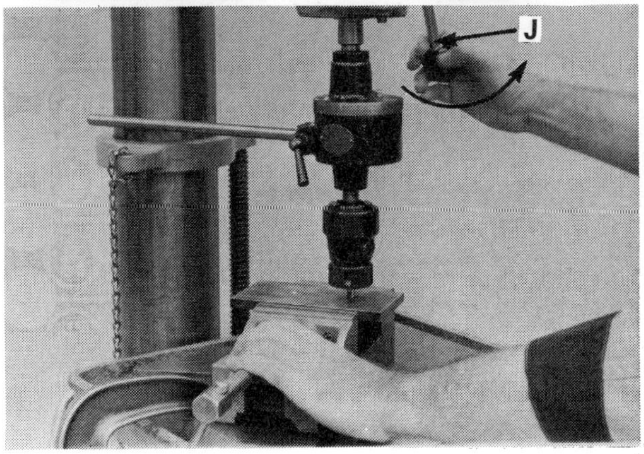

8.
- When the tap is through the work, release the down pressure on the feed handle (J).
- Hold the vise down firmly on the table and against the table stop for taps up to $\frac{5}{16}$" (8 mm) diameter. For taps over $\frac{5}{16}$" (8 mm), the vise must be clamped to the table.
- The tap will back out of the hole as long as up pressure is applied on the handle (J).

CENTER DRILLING

Work which is to be machined between lathe centers must have a proper center hole drilled in each end for the lathe centers to enter. The hole also provides a good bearing surface for the workpiece.

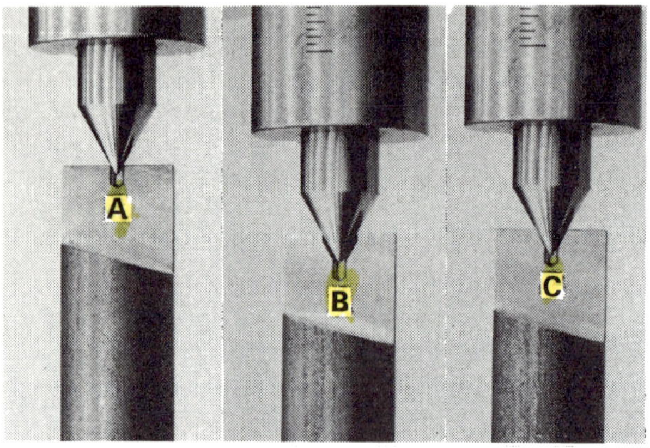

1. • Hole "A" is too shallow and will result in damage to both the lathe center and the work.
   • Hole "B" is too deep and does not allow the taper on the lathe center to contact the taper of the center hole.
   • Hole "C" provides a proper bearing surface and support for the work.

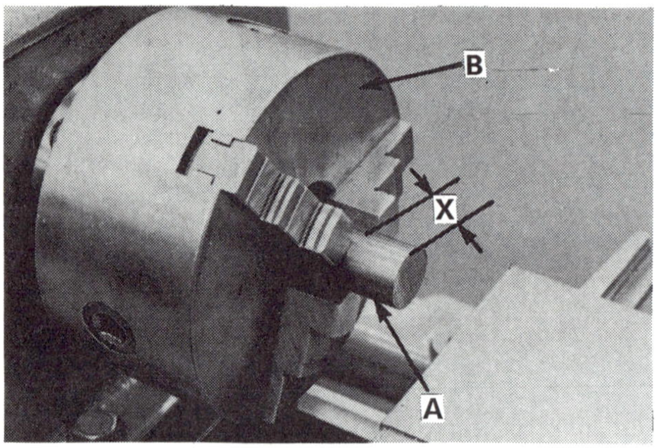

2. • Mount the round workpiece (A) in a three-jaw universal chuck (B).
   • The workpiece should be held short and never extend more than three times the work diameter beyond the chuck jaws (distance X).

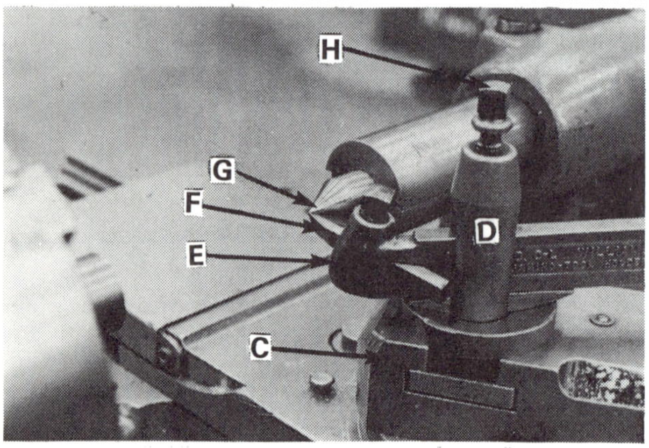

3. • Swing the compound rest (C) approximately 30° to the right.
   • Move the toolpost (D) to the left side of the compound rest.
   • Hold the toolholder (E) short in the toolpost.
   • Mount a facing tool (F) in the toolholder .
   • Set the point of the facing tool to the same height as the tailstock center point (G) and tighten the toolpost screw (H) slightly.

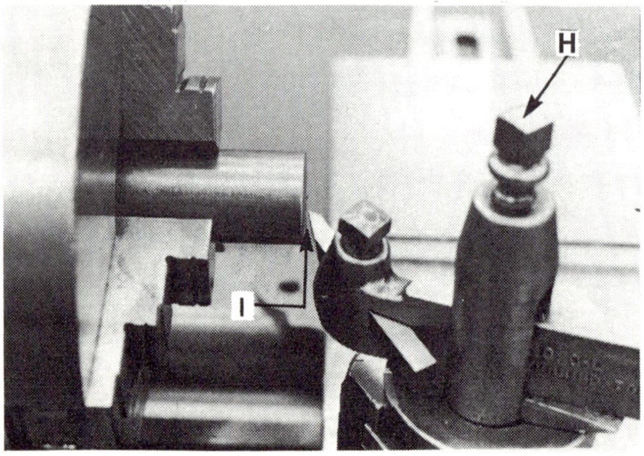

4. • Swivel the toolholder so that the point of the facing tool is pointing to the left and a space (I) is left between the side of the toolbit and the end of the workpiece.
   • Tighten the toolpost screw (H) securely.

5. • Set the lathe for the proper spindle speed (approximately 400 *rpm* for a 1" diameter).
   • Using the carriage handwheel (J) bring the facing tool close to the center of the workpiece.
   • Start a small cut and then lock the carriage in position by tightening the carriage lock (K) using two finger pressure on the wrench.

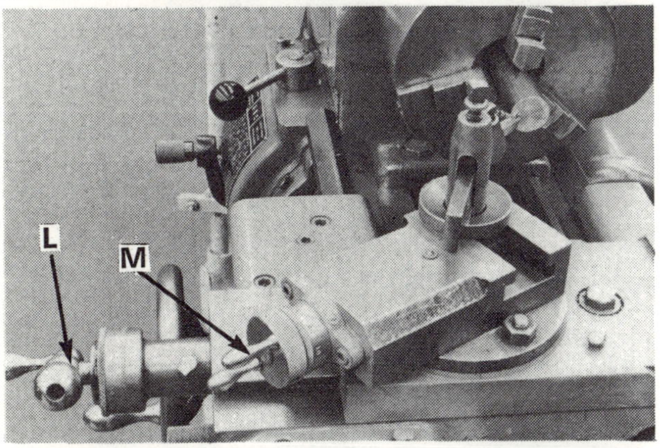

6. • Use the crossfeed handle (L) and feed the cutting tool from the center out.
   • Return the cutting tool to center if the surface is not cleaned up.
   • Set the depth of cut by feeding the compound rest handle (M).
   • Take successive cuts until the end surface is square.

7. • Clean the tailstock spindle and the tapered shank of the drill chuck.
   • Mount the drill chuck (N) into the tailstock spindle (O).
   • Fasten the proper size center drill (P) in the drill chuck. See the Table in Frame #11 for center drill sizes.

8. • Check to see that the lines (Q) on the top and bottom half of the tailstock are in line.
   • If they are not in line, see Lathe Operation #2 on "Alignment of Centers" and correct the tailstock misalignment.

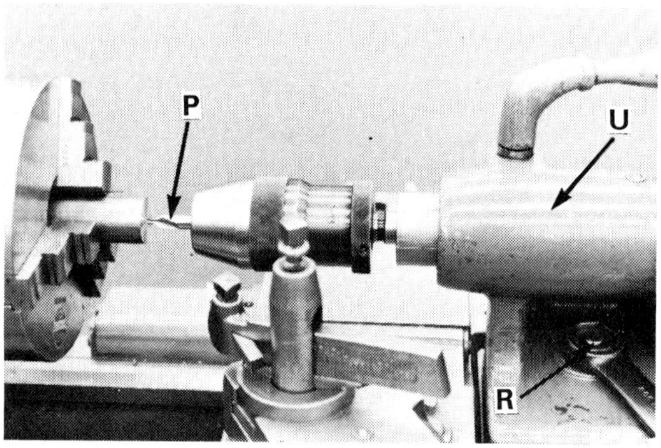

9. • Slide the tailstock (U) forward until the center drill (P) is close to the work.
   • Lock the tailstock in position by tightening the tailstock clamp nut (R).

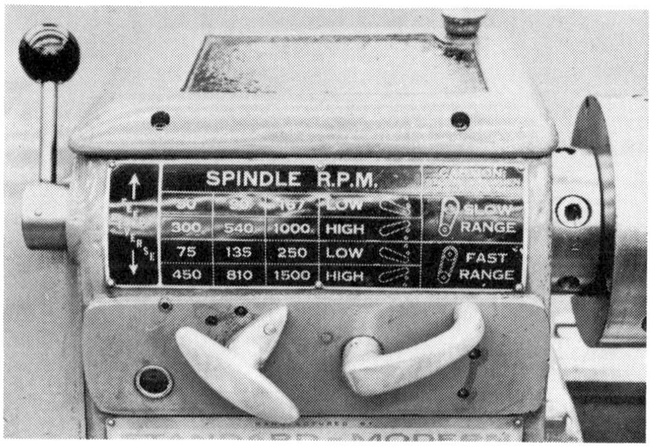

10. • Use the formula: $rpm = \frac{CS \times 4}{Dia.}$ to calculate the spindle speed for center drilling.
    • Use the diameter at the top of the center hole when calculating the *rpm*.
    • Set the spindle speed to approximately ¾ of the calculated *rpm*.

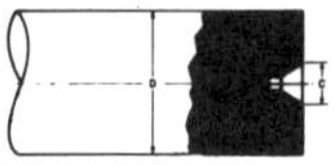

Size	Work Diameter D	Diameter of Countersink C	Drill Point Diameter	Body Size
1	3/16 to 5/16	3/32	3/64	1/8
2	3/8 to 1/2	9/64	5/64	3/16
3	5/8 to 3/4	3/16	7/64	1/4
4	1 to 1 1/2	15/64	1/8	5/16
5	2 to 3	21/64	3/16	7/16
6	3 to 4	3/8	7/32	1/2
7	4 to 5	15/32	1/4	5/8
8	6 and over	9/16	5/16	3/4

11. • Use the Center Drill Size table and
    (a) Locate the work diameter (D) in the second column.
    (b) Find the size center drill which should be used for that diameter in the left hand column.
    (c) Determine how deep the countersink portion should enter the work in the third column (C).

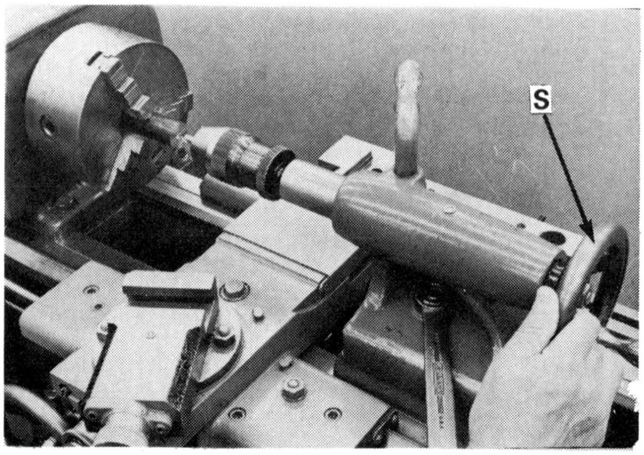

12. • Start the lathe spindle revolving.
    • By turning the tailstock handwheel (S), slowly feed the center drill into the work.
    • Occasionally apply cutting fluid to aid in the cutting action and produce a smooth hole.
    • Drill until the front face of the center hole is to the correct diameter.

ALIGNMENT OF LATHE CENTERS

A parallel diameter can only be machined on work mounted between centers if the headstock and tailstock centers are in line with each other. If these centers are out of alignment, a taper (one end larger than the other) will be cut.

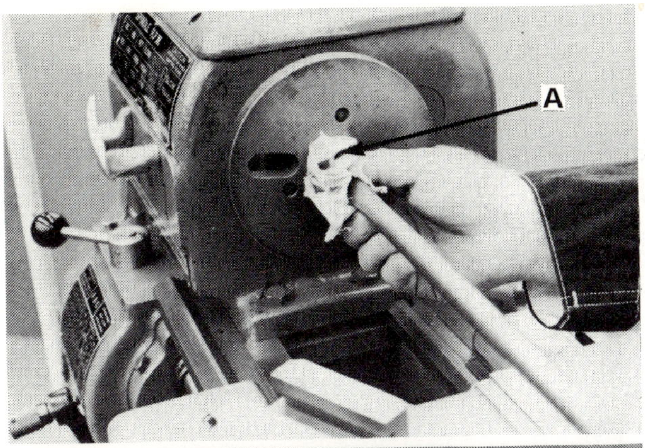

1. • Thoroughly clean the tapered hole in the head-stock spindle (A) with a cloth wrapped around a stick.
   • Do not start the lathe for this operation.
   • Remove all burrs or nicks from the spindle hole with a half round file or scraper.

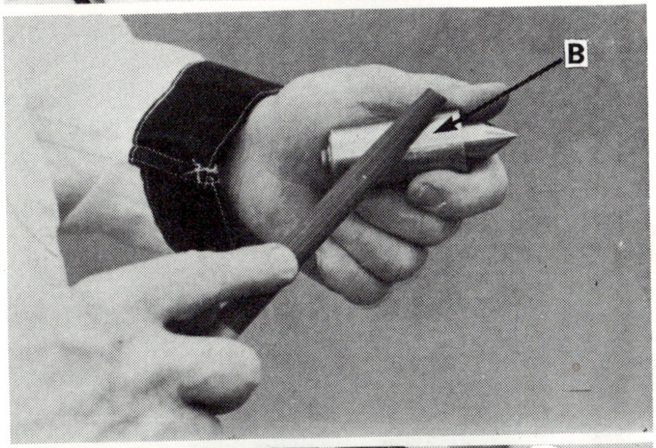

2. • Remove all burrs from the spindle sleeve (B) with a file or oil stone.
   • With *sharp snap*, insert the spindle sleeve *firmly* into the headstock spindle.
   • Clean the tailstock center and mount it in the tailstock.

3. • Check to see that the registration lines (C and D) on the top and bottom half of the tailstock are in line.
   • The top line (C) offset to the right will result in a work's diameter being cut smaller towards the headstock.

4. • Check the registration lines on the tailstock for alignment.
   • The top line (C) offset to the left will result in a work's diameter being larger towards the headstock.

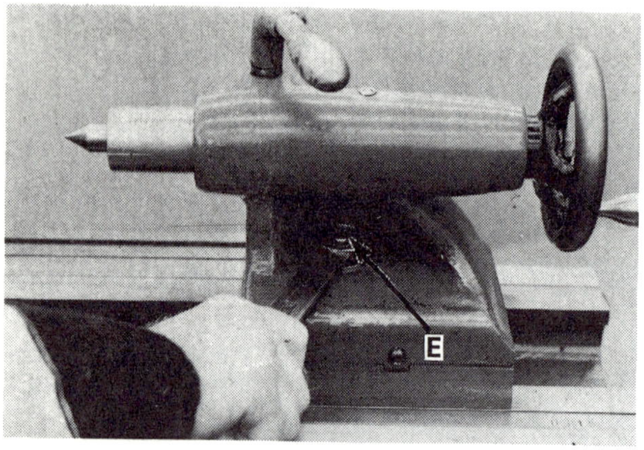

5. • If the registration lines are not in alignment, the tailstock must be adjusted.
   • Loosen the tailstock clamp nut (E).

6. • Loosen one adjusting screw (F) and tighten the opposite one until the registration lines are aligned.
   • Tighten the adjusting screw that was loosened.
   • Check to see that the tailstock lines are in exact alignment and correct if necessary.
   • Tighten the tailstock clamp nut.

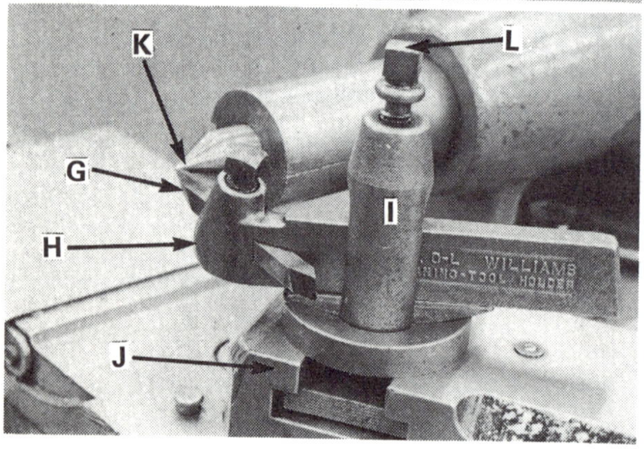

7. TRIAL CUT METHOD
   • Hold a turning toolbit (G) short in the toolholder (H).
   • Extend the toolholder approximately the width of a thumb past the toolpost (I).
   • Move the toolpost to the left-hand side of the compound rest (J).
   • Set the point of the toolbit even with the lathe center point (K) and tighten the toolpost screw (L) securely.

8. • Set the lathe for the proper speed, approximately 400 *rpm* for a 1″ (25.4 mm) diameter, at a fine feed.
   • Make a light trial cut (until the diameter cleans up) about ¼″ (6.35 mm) along the length of the stock.
   • Stop the feed and note the reading on the crossfeed micrometer collar (M).
   • Turn the crossfeed handle counterclockwise to bring the toolbit away from the work.

9. • Turn the carriage handwheel to bring the toolbit close to the headstock end.
   • Turn the crossfeed handle until it is at the same micrometer collar setting as the first cut.
   • Take a trial cut of about ½" (12.7 mm).
   • Stop the lathe and bring the toolbit away from the work.

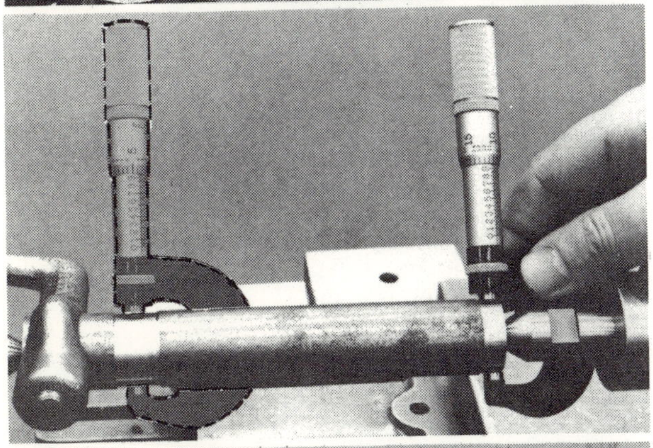

10. • Measure both diameters with a micrometer.
    • If both diameters are not the same, it will be necessary to adjust the tailstock as indicated in Frame #6.
    • Repeat Frames 7 to 9 until both diameters are the same.

11. TEST BAR AND DIAL INDICATOR METHOD
    • Clean both lathe centers and the center holes in the test bar (N).
    • Mount the test bar between centers so that there is no end play (movement) on the test bar.
    • Fasten a dial indicator (O) on the lathe and set the indicator plunger (P) on the center line of the bar.
    • Turn the crossfeed handle until the indicator registers approximately ¼ of a turn on the diameter at the tailstock end.

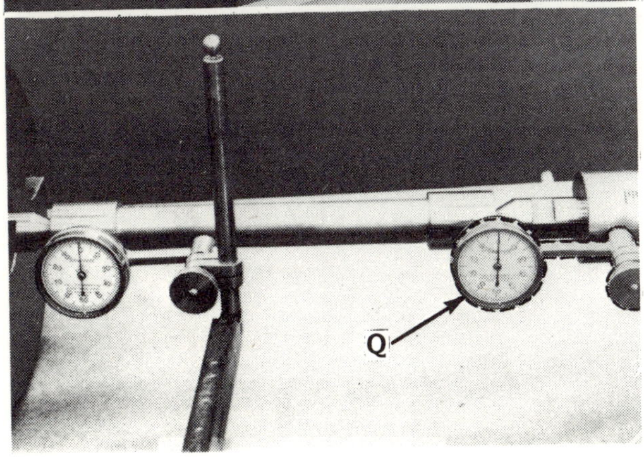

12. • Set the indicator bezel (Q) to zero.
    • Turn the carriage handwheel to bring the indicator plunger on the diameter at the headstock end.
    • Note the indicator reading; if both readings are the same, the centers are in line.
    • If the two readings are different, see Frame #6 and adjust the tailstock until both indicator readings are the same.

MOUNTING WORK BETWEEN CENTERS

When machining work between lathe centers, it is important that the workpiece and the cutting tool be set up properly, otherwise the tailstock center and the workpiece can be damaged. This could result in the work's being thrown out of the lathe and causing injury to the operator.

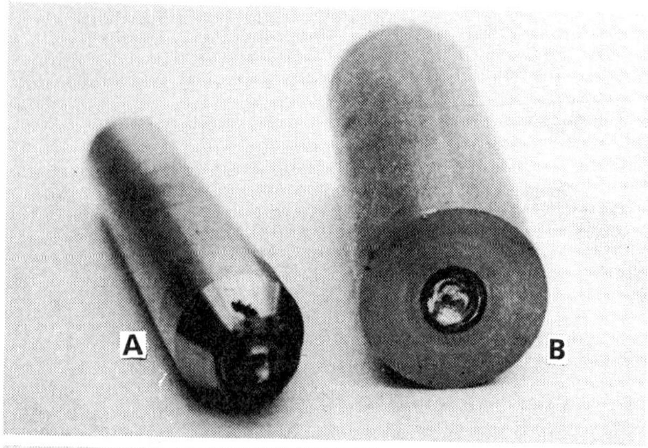

1. Improper center tension may result in the following conditions:
   - Figure A shows a lathe center whose point has been burned off due to excessive heat and friction.
   - Figure B shows the center point "welded" into the center hole of the workpiece.
   - These conditions can be avoided if the work is properly mounted between the lathe centers.

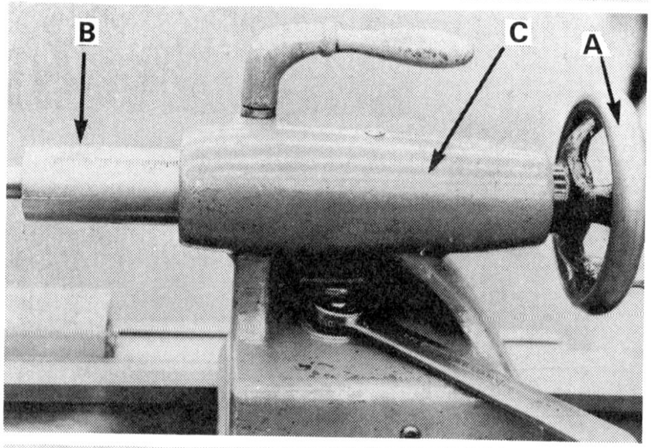

2. - Clean and align the lathe centers.
   - Turn the tailstock handwheel (A) until the tailstock spindle (B) extends 2½" to 3" (63.5 to 76.2 mm) beyond the tailstock (C).
   - This will allow clearance between the carriage and tailstock so that the cutting tool can reach the end of the workpiece.

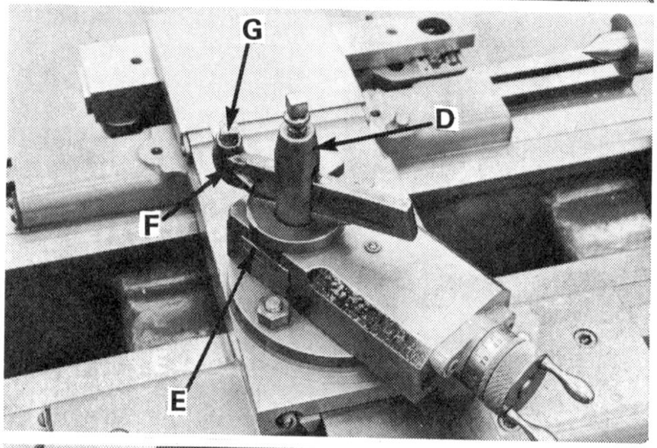

3. - Move the toolpost (D) to the left side of the compound rest (E).
   - Position the toolholder (F) short (about the width of a thumb) between the toolholder screw (G) and the toolpost.

4. - Mount a turning toolbit (H) in the toolholder and have it extending about ½" (12.7 mm).
   - Set the point of the toolbit even with the lathe center point (I).
   - Tighten the toolpost screw (J) securely.

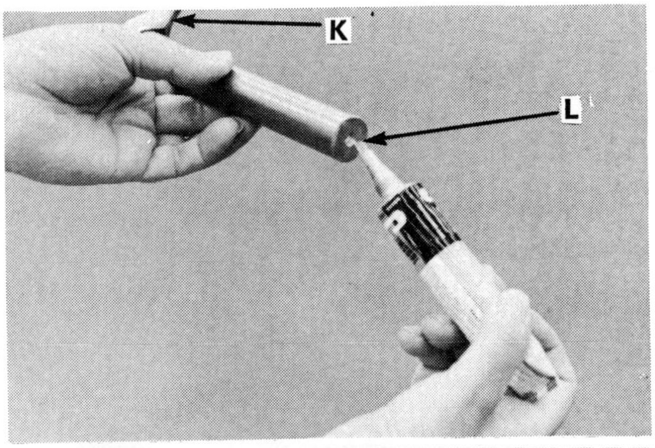

5. • Place a lathe dog (K) on the left-hand end of the work.
   • Apply a suitable lubricant to the right-hand center hole (L) of the work.

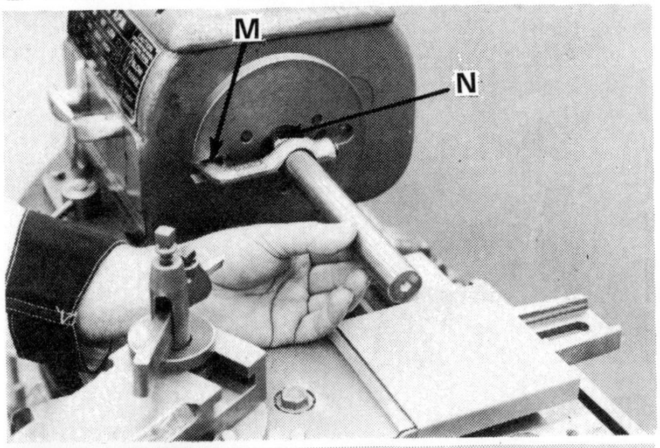

6. • Turn the headstock spindle until a slot in the drive plate (M) is in a horizontal position.
   • Place the work on the live center (N) of the lathe.

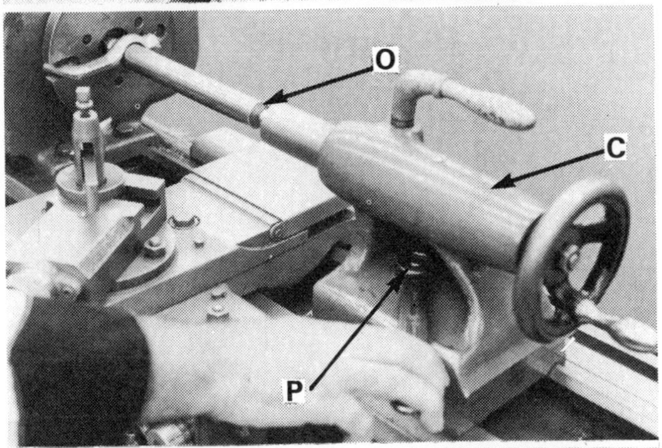

7. • Slide the tailstock (C) up until the tailstock center (O) enters the center hole of the work-piece.
   • Tighten the tailstock clamp nut (P) to fasten the tailstock in position.

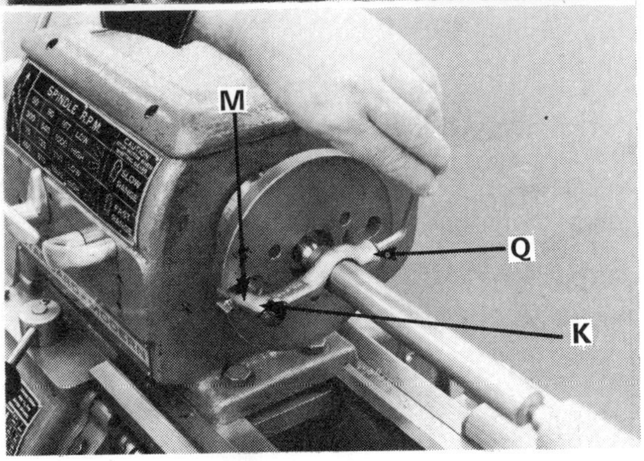

8. • Adjust the tail of the lathe dog (K) into the drive plate slot (M).
   • Be sure that the lathe dog is not binding in the drive plate slot.
   • Tighten the lathe dog screw (Q) securely onto the workpiece.

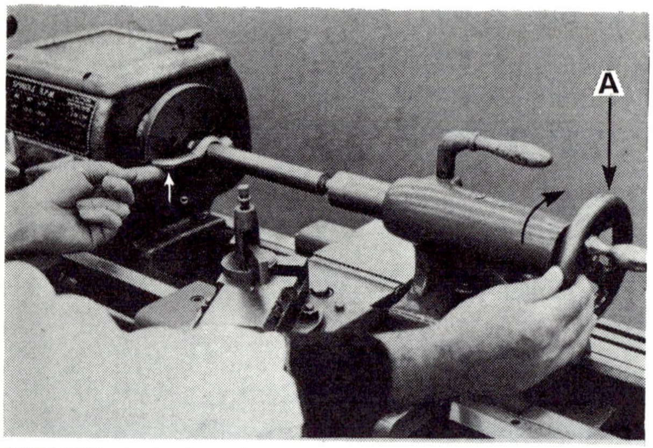

9. • Hold the tail of the lathe dog up in the drive plate slot.
   • Turn the tailstock handwheel (A) clockwise until it just holds the dog in the up position.

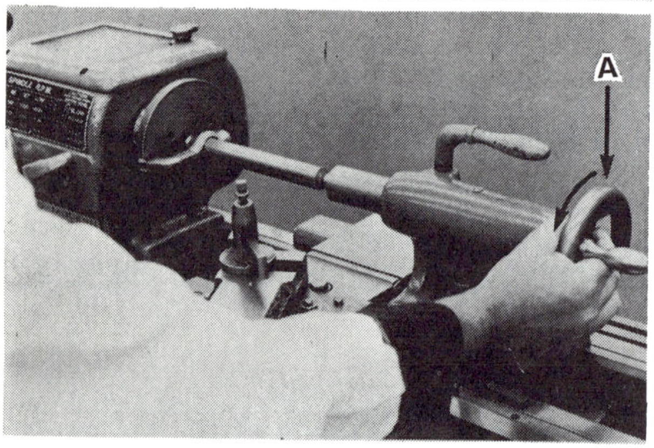

10. • Turn the tailstock handwheel (A) counter-clockwise until the tail of the dog drops in the drive plate slot.
    • Hold the handwheel in this position so that the center adjustment does not change.

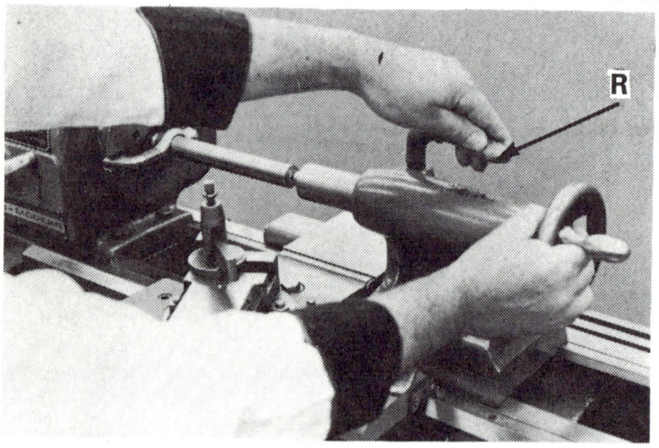

11. • With the left hand, tighten the tailstock spindle clamp (R).
    • This will lock the center adjustment in the proper position.

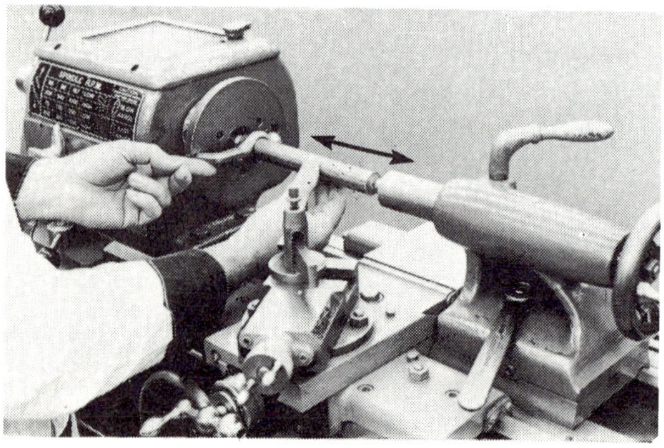

12. • Recheck the center adjustment of the work:
    (a) Hold the tail of the dog up in the slot and remove your hand. The tail of the dog should drop of its own weight.
    (b) Attempt to move the work endwise. There should be no end play between the work and the lathe centers.
    • Readjust the center tension if necessary.

FACING

Facing the ends of a workpiece may be performed with the work mounted between lathe centers or in a chuck. The purpose of facing is to: (a) smooth and square the ends, (b) provide an accurate surface for measurement, (c) cut the work to the correct length.

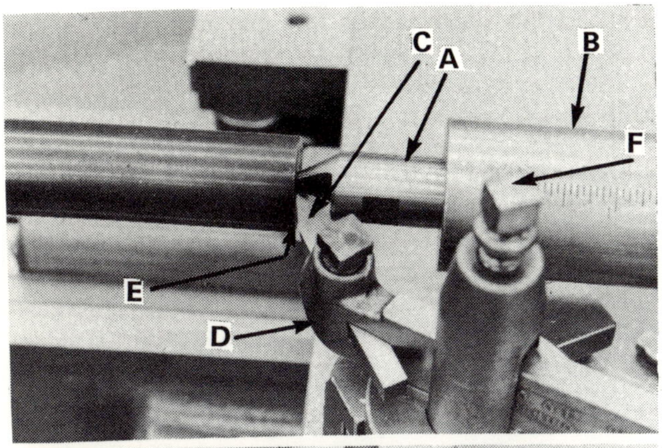

1.
   - **Make sure the lathe centers are aligned (operation #2).**
   - **Place a half center (A) in the tailstock spindle (B).**
   - **Mount a facing tool (C) in the toolholder (D) and set it to the lathe center point.**
   - **Mount the work between centers (operation #3).**
   - **Swivel the toolholder until the point of the facing tool is close to the work with a space (E) along the side.**
   - **Tighten the toolpost screw (F) securely.**

2.
   - **Have the lathe revolving at the correct speed for facing.** $\text{rpm} = \dfrac{CS \times 4}{\text{Dia.}}$
   - **Bring the facing tool as close to the lathe center as possible by turning the crossfeed handle (G).**
   - **Turn the carriage handwheel (H) until a light cut is started at the center of the work.**

3.
   - **Tighten the carriage lock (I) using two finger pressure on the wrench.**
   - **This will prevent the carriage (and facing tool) from moving during the cut and ensure that a square surface is cut.**

4.
   - **Use two hands and turn the crossfeed handle (J) counterclockwise slowly and steadily to bring the facing tool from the center to the outside of the work.**
   - **If the entire surface is not cleaned up, repeat Frames 3 and 4 until the entire surface is machined.**

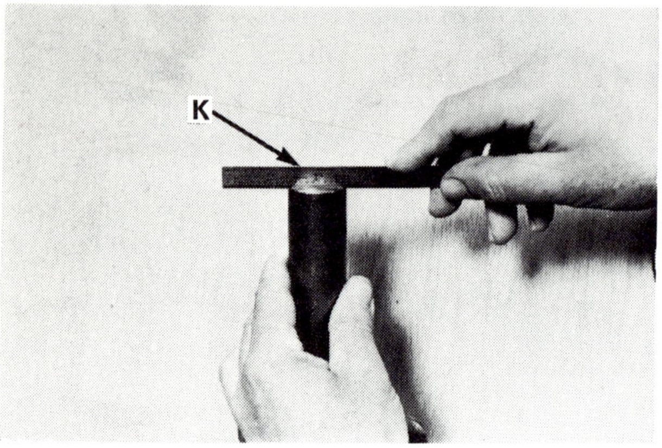

5. • With the work either in or removed from the lathe, check the faced surface for flatness with the edge of a rule (K).
   • If it is correct, no light should show between the rule and the faced surface.
   • If the center of the surface is *high*, recheck the setup to make certain that there is a space left between the point and side of the facing tool.
   • If the center is *low*, the carriage or tool setup has moved during the cut.
   • Face the other end until the work is the correct length.

6. • Work being faced in a chuck should not extend more than three times its diameter (X) beyond the chuck jaws.
   • This will prevent vibration or chattering and reduce the possibility of bending the work.
   • Tighten the work securely in the chuck.

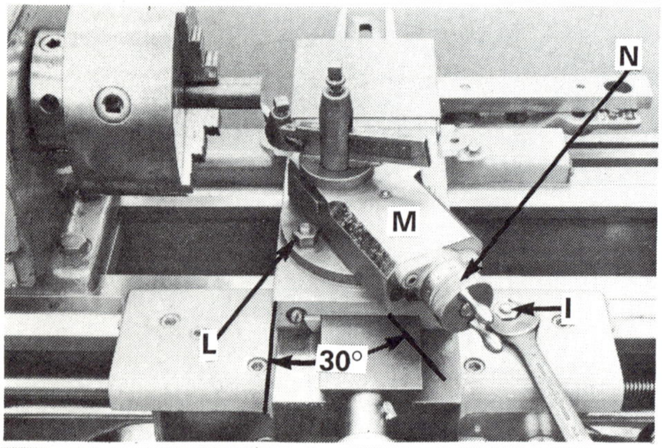

7. • Loosen the lock nuts (L) and swivel the compound rest (M) 30° to the right.
   • Tighten the lock nuts using two-finger pressure on the wrench.
   • Set up the facing tool the same as for facing between centers.
   • An infeed of .040 of the compound rest screw (N) will move the tool sideways .020 or *always one-half the amount of compound rest feed.*
   • Tighten the carriage lock (I) and proceed with the facing operation.

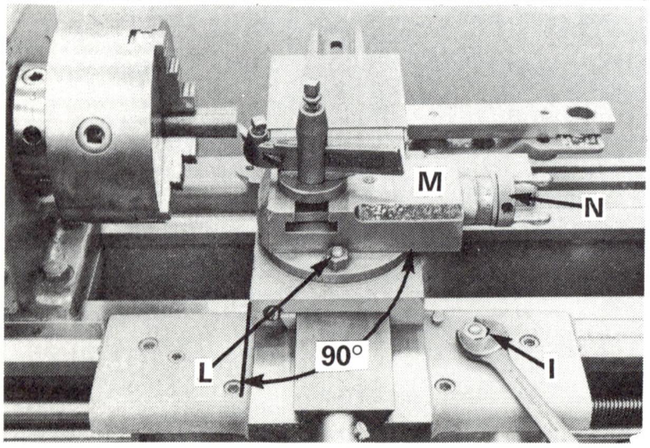

8. • When a series of short steps or shoulders must be accurately faced, set the compound rest (M) at 90° to the cross-slide.
   • Tighten the carriage lock (I).
   • Face the end until the surface is flat.
   • Measure the work length and face to length using the graduated collar of the compound rest feed screw (N).

TURNING TO AN ACCURATE DIAMETER

Turning to an accurate diameter is one of the most important operations on a lathe. If proper procedures are followed, accurate work will be produced.

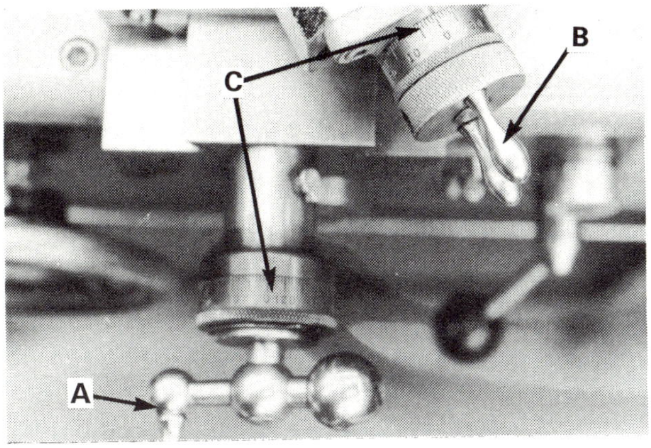

1. • The crossfeed handle (A) and compound rest handle (B) on a lathe are both equipped with micrometer graduated collars (C), with each line usually representing .001.
   • The cutting tool is set for a depth of cut by the crossfeed handle (A).

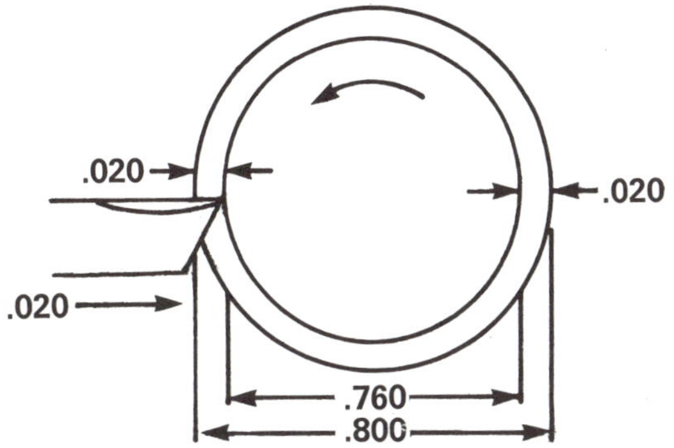

2. • Since the workpiece in a lathe revolves, material will be removed from the circumference of the workpiece.
   • If the crossfeed handle is turned clockwise .020, the cutting tool would be moved in .020 and the work diameter would be reduced by .040.
   • Therefore on machines *in which the work revolves,* the amount of material removed is always *twice the amount set on the graduated collar.*

3. • Calculate the speed at which to set the lathe for the diameter and type of workpiece material:

$$\text{rpm} = \frac{\text{CS} \times 4}{\text{Dia.}}$$

   • EXAMPLE: To cut a 1″ diameter piece of machine steel (100 CS)

$$= \frac{100 \times 4}{1}$$

$$= 400 \text{ rpm}$$

   • Set the headstock levers (D and E) to the correct positions.

4. • Set the levers to the positions indicated on the quick-change gearbox (F) for the desired feed rate.
   (a) *Roughing cuts*, used to remove metal quickly, should be from .007 to .015 for each revolution of the workpiece.
   (b) *Finishing cuts*, used to cut work to size and provide a good surface finish, should be from .003 to .005.

5. • Mark the length of the work to be machined by:
   (a) Holding a rule against the work, move the carriage until the cutting tool point is at the desired length, and then cut a light groove on the work.

   OR

   (b) By placing a center punch mark at the correct length before the work is mounted in the lathe.

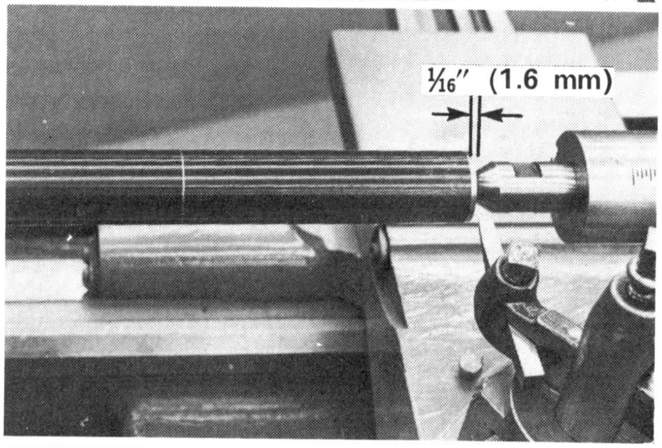

6. • Move the carriage until the cutting tool overlaps the right-hand end of the workpiece by approximately 1/16″ (1.6 mm).
   • While the lathe is revolving, feed the cutting tool into the work with the crossfeed handle until a light cut is made *around the entire circumference* of the work.
   • *Do not move the crossfeed handle setting*; bring the carriage to the right until the cutting tool clears the end of the workpiece.

7. • Turn the crossfeed handle (A) *clockwise* .010 and take a trial cut 1/4″ (6.35 mm) long at the end of the work.
   • Disengage the feed, stop the lathe, *do not move the crossfeed handle setting.*

8. • Move the carriage to the right until the cutting tool clears the end of the work.
   • Measure the diameter of the trial cut with a micrometer.

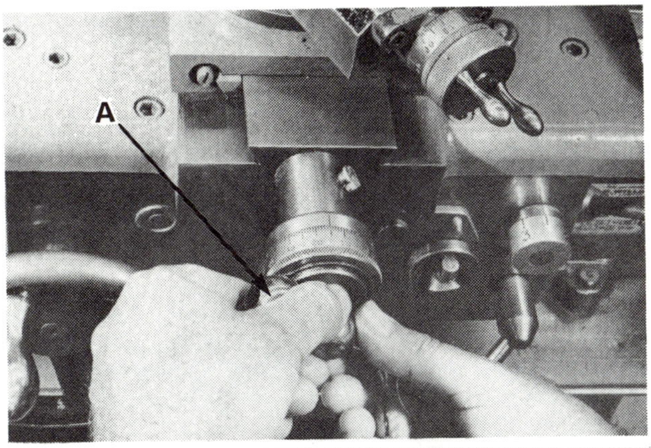

9.
- **Set the crossfeed handle (A) for the rough cut (.030" or .8 mm larger than the finish size of the work).**
- Take a trial cut ¼" (6.35 mm) long and measure the diameter. *Do not move the crossfeed handle setting.*

10.
- Reset the cutting tool if necessary and take a rough cut.
- Check the center tension about every 3" (76.2 mm) of length by disengaging the feed, stopping the lathe, *but do not move the crossfeed handle setting.*
- Move the cutting tool approximately 1/16" (1.6 mm) to the right and adjust the center tension if necessary.

11.
- Start the lathe, engage the feed and continue with the rough cut.
- Disengage the automatic feed about 1/16" (1.6 mm) from the desired length.
- Feed the cutting tool to the correct length by moving the carriage by hand.

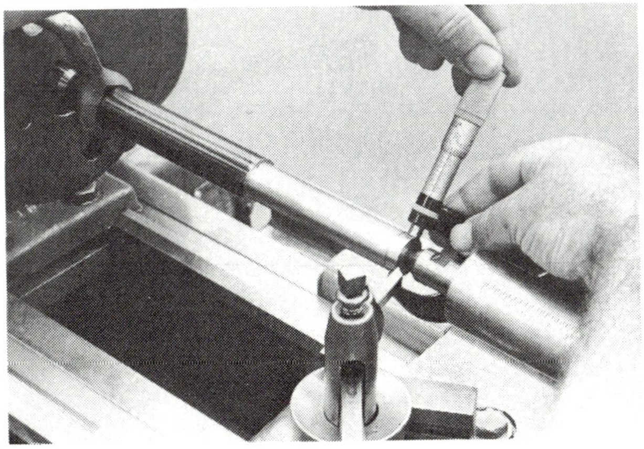

12.
- *Do not touch the crossfeed handle setting.*
- Move the carriage to the right until the cutting tool is just past the end of the work.
- Set the lathe for a .003 to .005 feed.
- Turn the crossfeed handle clockwise .005, take a trial cut for ¼" (6.35 mm) long and measure this diameter.
- Set the cutting tool to the required depth and take the finish cut.

TURNING A SHOULDER

Whenever more than one diameter is machined on a shaft, the section separating each diameter is called a shoulder or step. Square, fillet, and undercut shoulders are most commonly used in machine shop work.

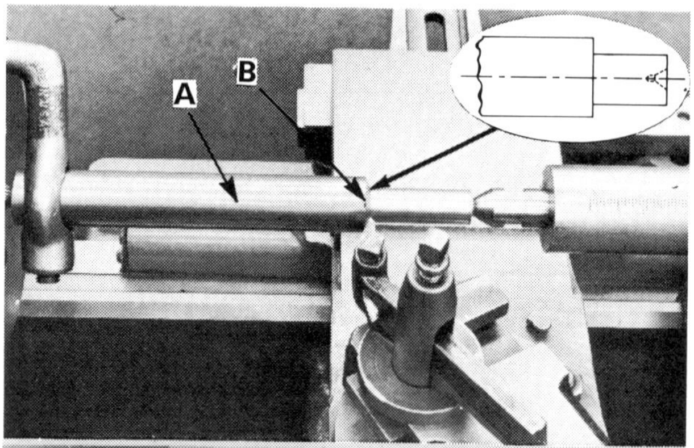

SQUARE SHOULDER

1. • Turn the large diameter (A) to size.
   • Mark the length of the small diameter with a center punch mark (B) or by cutting a light groove with a toolbit.
   • Rough and finish turn the smaller diameter to size and to within 1/32" (.80 mm) of the required length.

2. • Insert a facing tool (C) and swivel the toolholder (D) until the point of the tool is set for facing.
   • Tighten the toolpost screw (E) securely.
   • Mark the small diameter, near the shoulder, with chalk or layout dye (F).
   • Start the lathe and feed the cutting tool with the crossfeed handle (G) until it lightly marks the chalk or layout dye.
   • Note the reading on the crossfeed graduated collar (H).

3. • Turn the crossfeed handle counterclockwise ½ turn and bring it back to within .005 of the original setting.
   • Start the lathe and turn the carriage handwheel (I) to start a light cut.
   • Turn the crossfeed handle (G) counterclockwise to face the shoulder.
   • Repeat this procedure until the shoulder is cut to the proper length.
   • Turn the crossfeed handle to the original graduated collar setting and take the final cut.

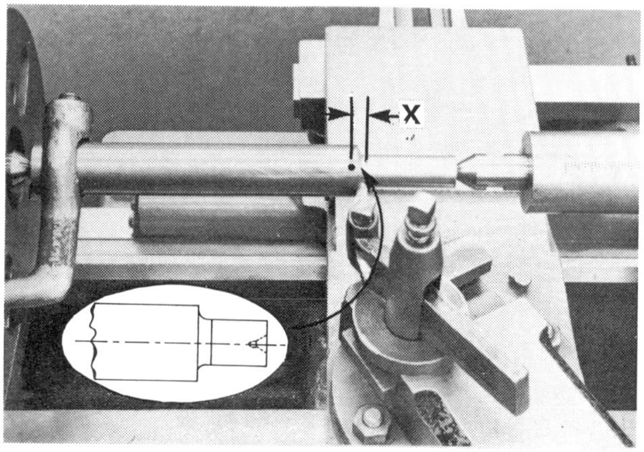

4. • FILLETED SHOULDER
   • Mark the length of the shoulder with a center punch mark or by cutting a light groove with a toolbit.
   • Rough and finish turn the small diameter shorter than the finished length.
   • (Allow enough material (X) for the size of radius to be cut.)

   EXAMPLE: A 3" (76.2 mm) length with a ⅛" (3.17 mm) radius should be turned 2⅞" (73 mm) long.

5. ● Mount a radius toolbit (J) and set it to center.
   ● Mark the small diameter, near the shoulder with chalk or layout dye (F).
   ● Start the lathe and bring the radius tool in until it lightly marks the chalk or layout dye.
   ● Note the reading on the crossfeed graduated collar (H).

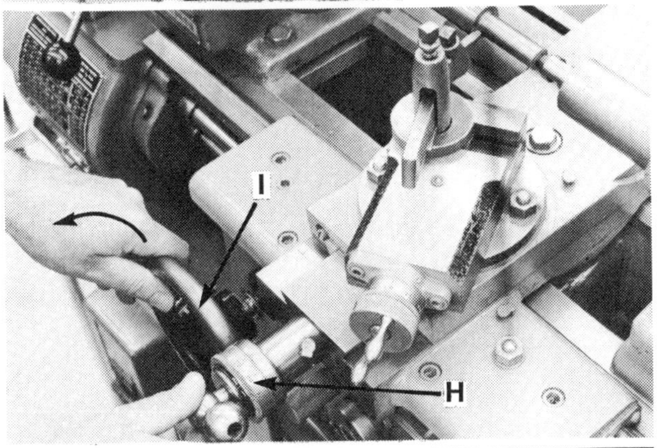

6. ● Turn the crossfeed handle counterclockwise ½ turn and then clockwise to within .005 of the original setting.
   ● Set the lathe to revolve at approximately ½ the speed used for turning.
   ● *Slowly* feed the tool to the correct length by turning the carriage handwheel (I).
   ● For the final cut, turn the crossfeed handle to the original graduated collar setting (H).

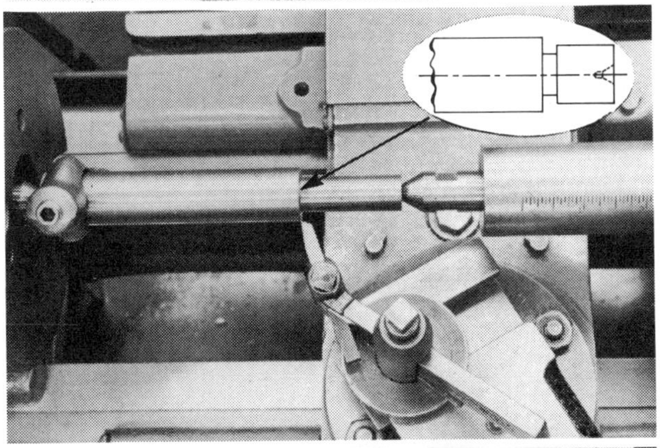

7. UNDERCUT SHOULDER
   ● Layout, rough and finish turn the small diameter to within 1/32" (.80 mm) of the correct length.
   ● Set up a facing tool and square the shoulder as in Frames #2 and 3.

8. Mount a suitable grooving tool (K) on center and at 90° to the work.
   ● Start the lathe and turn the carriage handwheel (I) until the side of the tool just touches the shoulder.
   ● Turn the crossfeed handle (G) until the tool just touches the diameter.
   ● Set the crossfeed graduated collar (H) to zero without moving the crossfeed handle.
   ● Apply cutting fluid while feeding the tool to the desired depth.

FILING AND POLISHING

Filing is used to remove burrs and sharp edges, reduce the diameter and improve the surface finish. Filing on a lathe should be kept to a minimum since too much filing will tend to produce a diameter which is out-of-round.

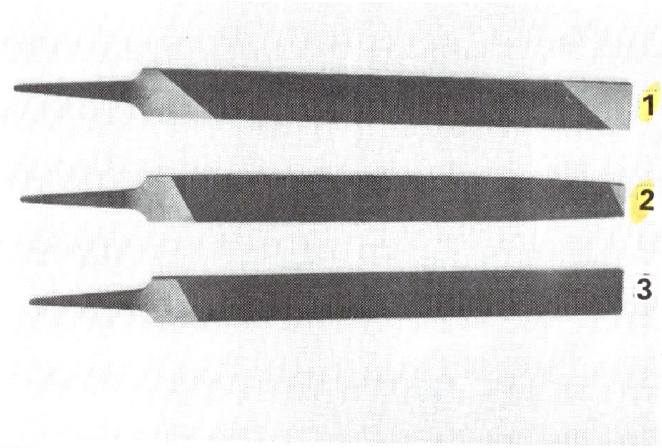

1. • The most common files used on the lathe are:
   (1) *The long angle lathe file* which provides rapid clearing of chips and reduces the possibility of tearing the metal.
   (2) *The mill file* which may be used for general-purpose filing on the lathe.
   (3) *The double cut file* which is used for the rapid removal of metal.

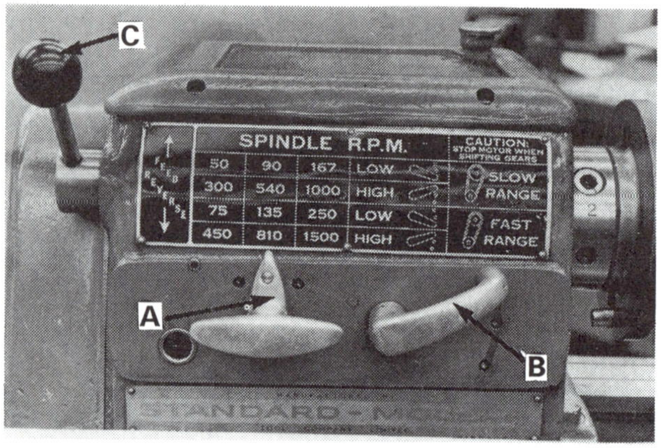

2. • Set the headstock levers (A and B) so that the lathe spindle revolves at approximately twice the turning speed.
   • Disengage the leadscrew and feed rod by placing the feed reverse lever (C) in the neutral position.

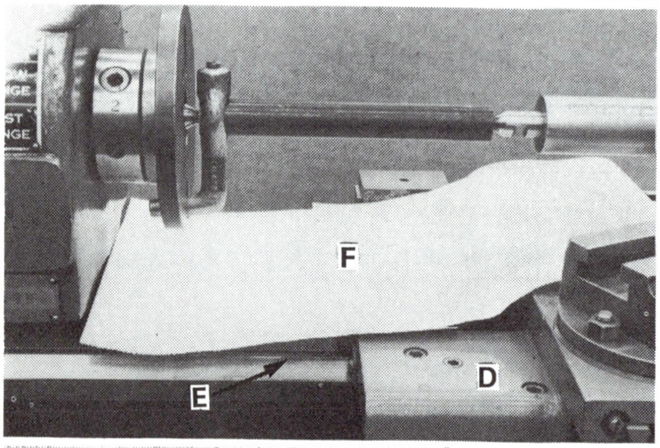

3. • Move the carriage (D) as far to the right as possible.
   • Cover the lathe bed (E) with paper (F) to prevent filings from damaging the lathe slides. DO NOT USE A CLOTH.

4. • Select the proper file for the job (single cut for finishing; double cut for roughing).
   • Be sure that the file is fitted with a secure handle (G). NEVER USE A FILE WITHOUT A HANDLE.
   • Hold the tip of the file by the fingers of the right hand (H) and the handle by the left hand (I).
   • Hold the file at approximately elbow height and keep it level.

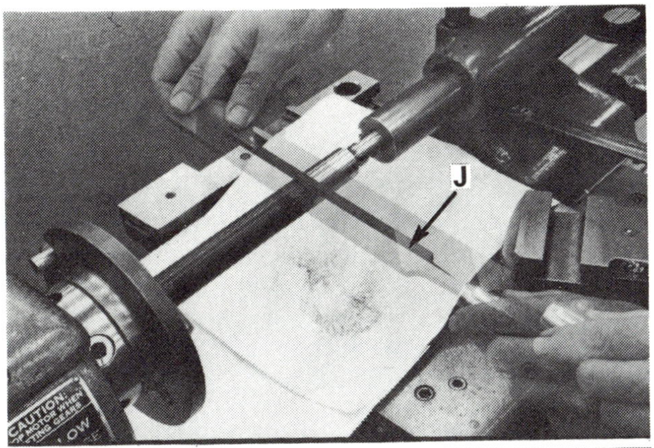

5. • Start the lathe spindle revolving.
   • Apply light down pressure (J) on the file and pass it over the diameter using a long forward stroke.
   • Allow the revolving work to return the file to the starting position. DO NOT LIFT OR APPLY PRESSURE ON THE FILE DURING THE RETURN STROKE.
   • Each pass of the file should overlap the previous pass by approximately one-half the file width.

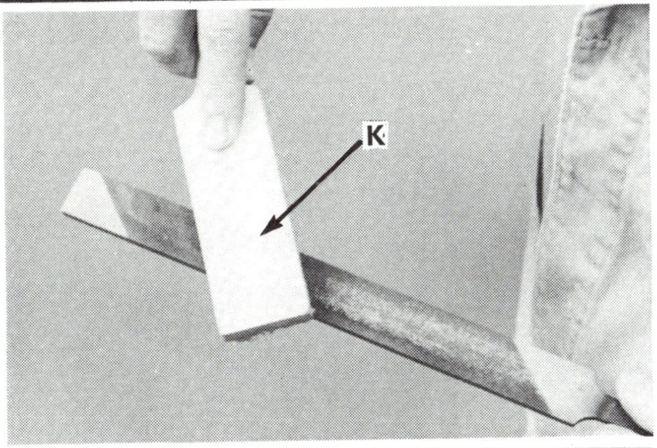

6. • Keep the file clean by frequently removing the chips with a file card (K) to prevent it from clogging and scratching the work surface.
   • Occasionally rub chalk on the file to prevent chips from clogging the teeth.

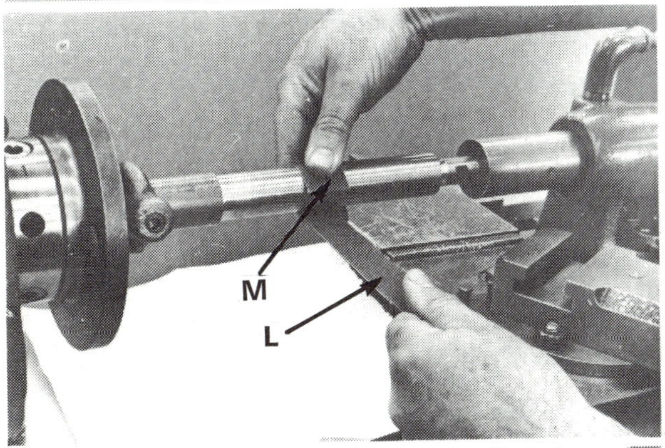

7. POLISHING
   • Select a piece of abrasive cloth 1″ (25.4 mm) wide by 10″ (254 mm) long to suit the material to be polished. (80 to 100 grit for roughing; 120 grit or finer for finishing.)
   • Set the lathe spindle at twice the filing speed.
   • Hold one end of the abrasive cloth (L) firmly in the left hand to prevent the other end from being drawn between the abrasive cloth and the workpiece.
   • With the fingers of the right hand, (M) press the abrasive cloth against the diameter.
   • Slowly move the abrasive cloth back and forth over the diameter.

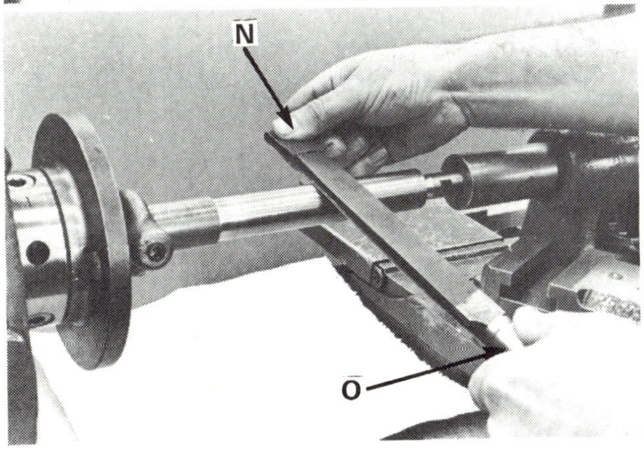

8. ALTERNATE METHOD
   • Fold one end of the abrasive cloth over the end of a file and hold it in position with the fingers of the right hand (N).
   • Hold the other end against the file handle with the left hand (O).
   • Polish the diameter using long passes similar to filing.

KNURLING

Knurling is the process of impressing a diamond-shaped or straight line pattern on round work to improve its appearance or provide a grip.

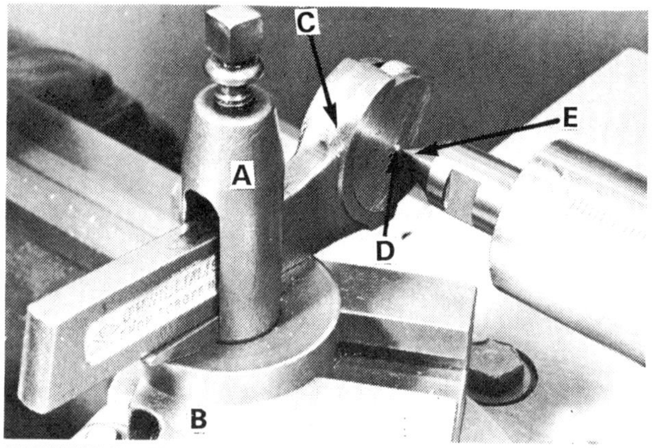

1. • **Move the toolpost (A) to the left side of the compound rest (B) and grip the knurling tool (C) short.**
   • **Set the center of the floating head (D), and not the knurling rolls, even with the lathe center point (E).**

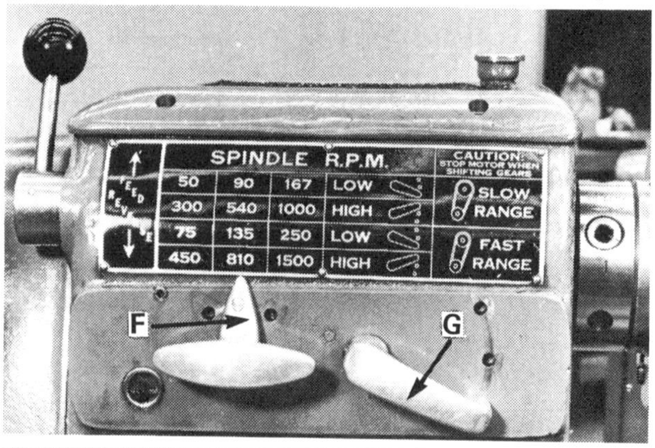

2. • **Set the spindle speed to approximately ¼ of the turning speed by setting the speed change levers (F and G).**
   (Use approximately 100 rpm for knurling a 1" diameter.)

3. • **Set the feed change levers (H) and (I) on the quick-change gearbox (J) for .020 to .030 feed. (Too slow a feed may ruin the knurling pattern, while too fast a feed will prevent the full knurl from forming.)**

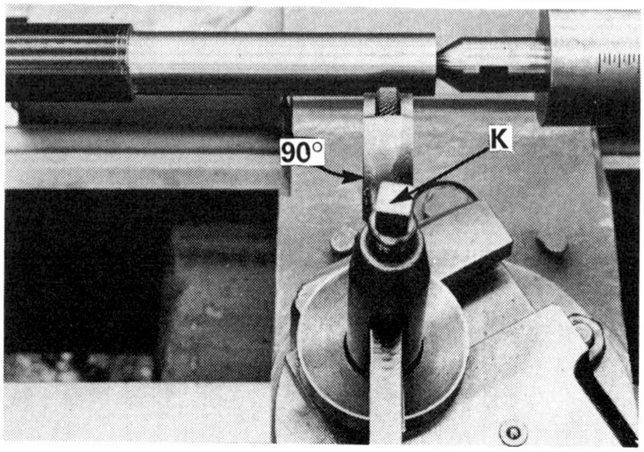

4. • **Mount the work between the lathe centers.**
   • **Adjust the center tension properly.**
   • **Set the knurling tool at 90° to the work.**
   • **Tighten the toolpost screw (K) securely.**

5. • Move the carriage so that one half of the knurling roll overlaps the work.
   • With the crossfeed handle, feed the knurling tool in until it contacts the work.
   • Force the knurling tool into the work .025 with the crossfeed handle and then start the lathe.

   OR

   • Start the lathe and then force the knurling tool in .025, or until the pattern forms.

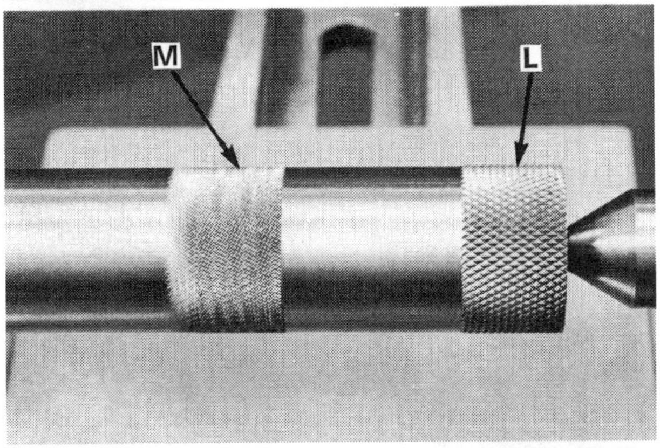

6. • Shut off the lathe and examine the knurled section. The knurl is correct if a true diamond pattern (L) is formed.
   • If the pattern is incorrect (M), recheck to see if the knurling tool is on center.
   • Lightly try the knurl at another spot on the work until the pattern is correct.

7. • Move the knurling tool to the end of the work and engage the automatic carriage feed.
   • Apply sufficient oil to the knurling rolls.
   • Never disengage the automatic feed during the knurling operation, otherwise rings (N) are formed on the work and the knurl is damaged.

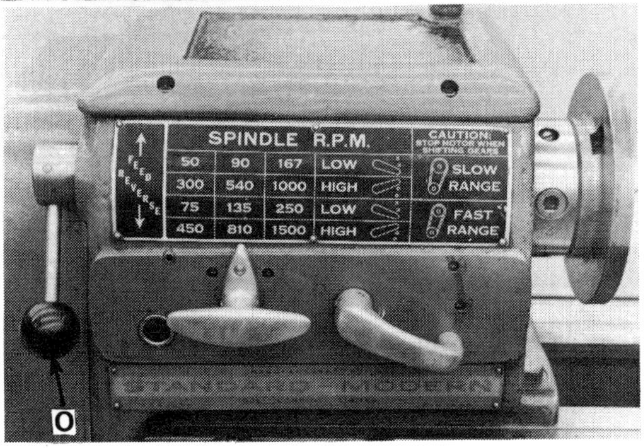

8. • When the knurling tool reaches the correct length to be knurled, shut off the lathe. *Do not disengage the automatic feed.*
   • If the diamond pattern is not to a point, place the feed reverse lever (O) in the down or reverse position.
   • With the feed reversed take another pass over the work until the diamond pattern is to a point.

A taper may be defined as a uniform increase or decrease in the diameter of a workpiece measured along its length. Tapers may be cut on a lathe by three methods: by offsetting the tailstock, with the taper attachment, or by using the compound rest.

1. **TAILSTOCK OFFSET METHOD**
   - **Calculate the amount the tailstock must be offset to cut the desired taper.**
   - **Loosen the tailstock clamp nut (A).**
   - **Loosen one tailstock adjusting screw and tighten the other (B and C) until the tailstock offset is correct.**
   - **Tighten the adjusting screw that was loosened and recheck the offset with a rule (D).**
   - **Correct the setting if necessary and then tighten the tailstock clamp nut.**

2. - **Set the cutting tool to center and mount the workpiece in the lathe.**
   - **Set the lathe speed and feed the same as for turning a diameter.**
   - **Rough turn the taper leaving it at least 1/16" (1.6 mm) oversize.**

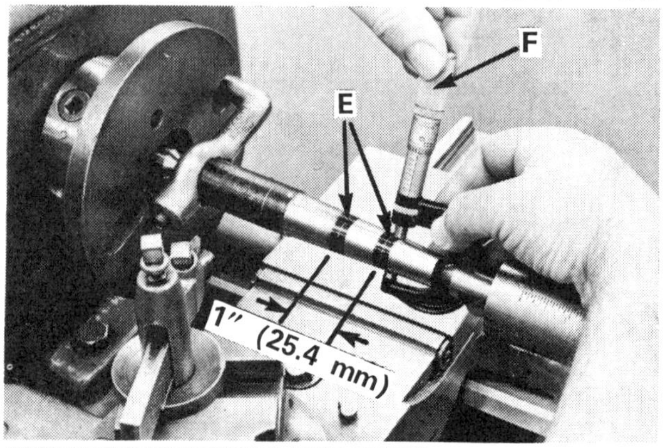

3. - **Divide the taper per foot by 12 to find the taper per inch.**
   - **Apply layout dye and mark off two lines (E) exactly 1" (25.4 mm) apart on the taper section.**
   - **Measure the diameter at these two lines with a micrometer (F).**
   - **Compare the difference in diameters with the amount of taper per inch; if these are the same, the taper is correct.**

4. - **If the taper is incorrect, mount a dial indicator (G) in the toolpost (H).**
   - **Set the indicator plunger on center and on one of the lines marked on the taper.**
   - **Loosen the clamp nut (A) and adjust the tailstock:**
     - **(a)** *towards the indicator* **if there is too little taper**
     - **(b)** *away from the indicator* **if there is too much taper.**
   - **Adjust the tailstock until the indicator reads 1/2 the error in 1" (25.4 mm) of the tapered length.**

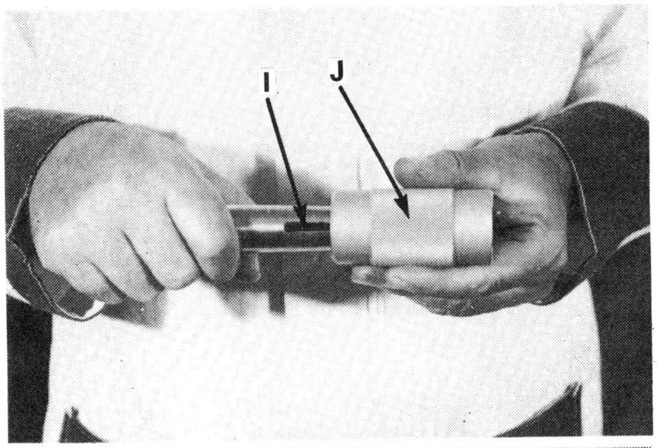

5. • Finish turn the taper.
   • With chalk or Prussian blue, mark three equally spaced lines (I) around the circumference, extending for the length of the taper.
   • Turn the part ¼ of a turn *counterclockwise* in a taper gage (J).
   • In the lathe file and fit the taper to the gage until the chalk or Prussian blue rubs off evenly.

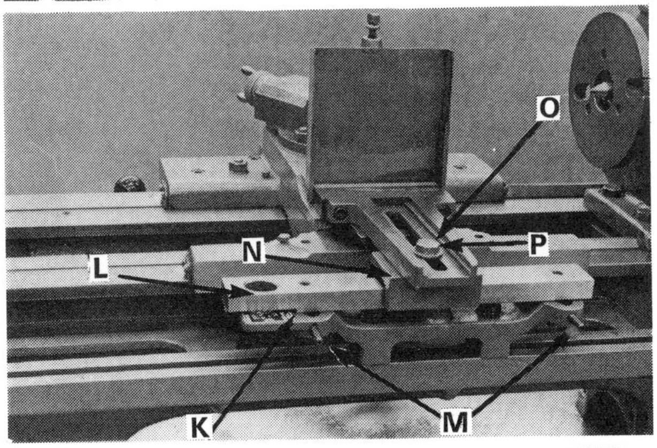

6. TAPER ATTACHMENT METHOD
   • *The plain taper attachment* consists of the following main parts:
   • The base plate (K), which is fastened to the lathe bracket.
   • A guide bar (L), which can be adjusted by the locking screws (M).
   • The sliding block (N), which is fastened to the cross-slide extension (O) by the binding screw (P).

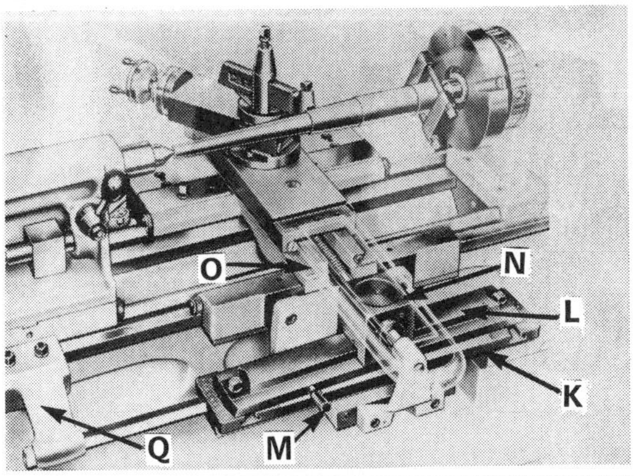

7. • *The telescopic taper attachment* consists of the following parts:
   • The base plate (K), which is fitted to the lathe saddle.
   • The guide bar (L), which can be adjusted to the desired taper by the locking screw (M).
   • The sliding block (N), which is connected to the cross-slide extension (O).
   • The clamping bracket (Q), which is locked onto the lathe bed.

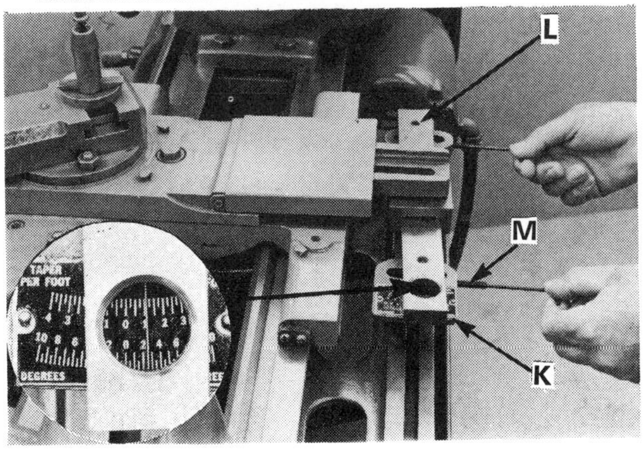

8. • To set a plain taper attachment, calculate the taper per foot or taper in degrees for the work to be cut.
   • Loosen the lock nuts holding the guide bar (L) to the base plate (K).
   • Adjust the locking screws (M) until the correct taper is indicated on the scale.
   • Tighten the guide bar lock nuts.

9. • If a correct taper gage (J) is available, mount it between centers.
   • Mount an indicator (G) with its plunger set to center and horizontal.
   • Use the binding screw (P) to connect the cross-slide extension (O) to the sliding block (N).
   • Adjust the taper attachment as in Frame #8 until there is no movement in the indicator as it is moved along the taper at which the gage (J) is set.

10. • Mount the work in the lathe.
    • Bring the cutting tool close to the workpiece.
    • Disconnect the cross-slide (R) from the cross-feed screw by removing the binding screw (P).
    • The cross-slide (and cutting tool) is now free to slide as guided by the taper attachment.

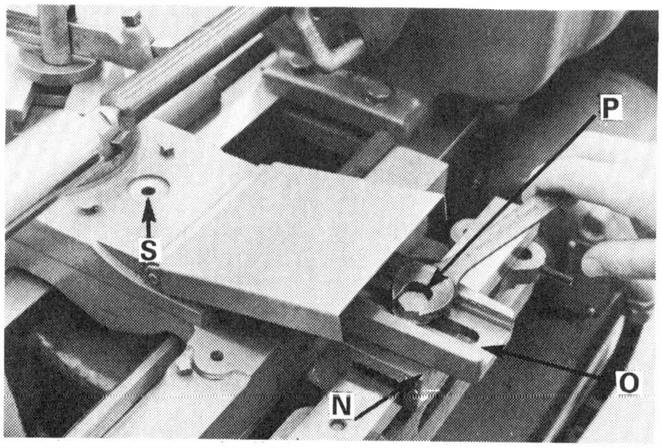

11. • Connect the cross-slide extension (O) to the sliding block (N) with the binding screw (P).
    • Tighten the binding screw using two-finger pressure on the wrench so that the threads will not be stripped.
    • Cover the cross-slide hole (S) with a suitable plug.

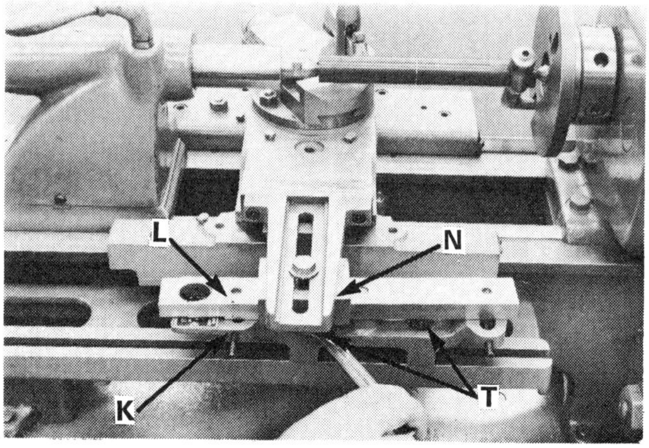

12. • Move the carriage so that the cutting tool is about in the center of the section to be tapered.
    • Loosen the lock nuts (T).
    • Position the base plate (K) so that the sliding block (N) is in the center of the guide bar (L).
    • Tighten the base plate lock nuts (T).
    • Cut and fit the taper as in Frames 4 and 5.

13. COMPOUND REST METHOD
- Check the drawing or blueprint to determine the amount of taper to be cut.
- For included angles, swivel the compound rest to one-half the angle.
- For angles indicated on one side only, set the compound rest to that angle.
- Loosen the compound rest lock nuts (U).

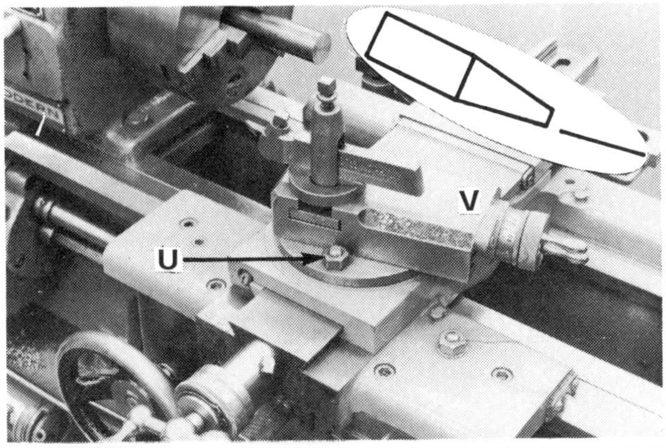

14.
- Swivel the compound rest (V) to the proper angle.
- Note the side of the compound rest and the angle on the front side of the taper should be parallel (see insert).
- Tighten both compound rest lock nuts (U) using only two-finger pressure on the wrench.

15.
- Swivel the toolholder (W) so that it is approximately 90° to the compound rest.
- Set the cutting tool (X) to center.
- Tighten the toolpost screw (Y) securely.
- Turn the compound rest handle (Z) counterclockwise to bring the compound rest (V) back enough so that the entire taper can be cut.

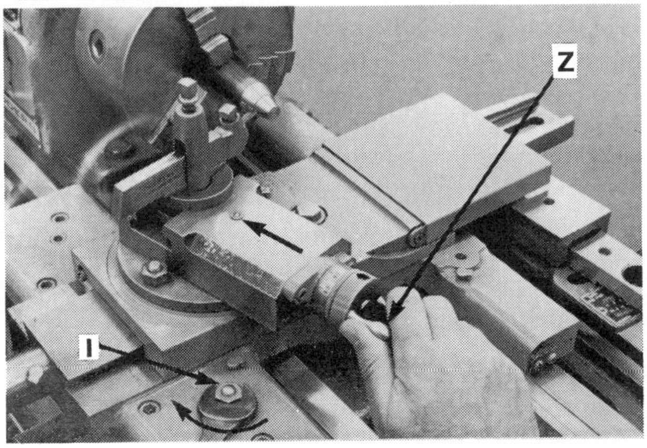

16.
- Bring the cutting tool close to the work by means of the carriage and crossfeed handles.
- Tighten the carriage lock (I).
- Set the depth of cut using the crossfeed handle.
- Cut the taper by turning the compound rest handle (Z).

A thread can be defined as a helical ridge of uniform section formed on the inside or outside of a cylinder or cone. Accurate threads, concentric with the work diameter, can be cut on a lathe by setting the leadscrew to revolve at a certain rate of speed. Since 60° thread is the most common, only this procedure will be explained.

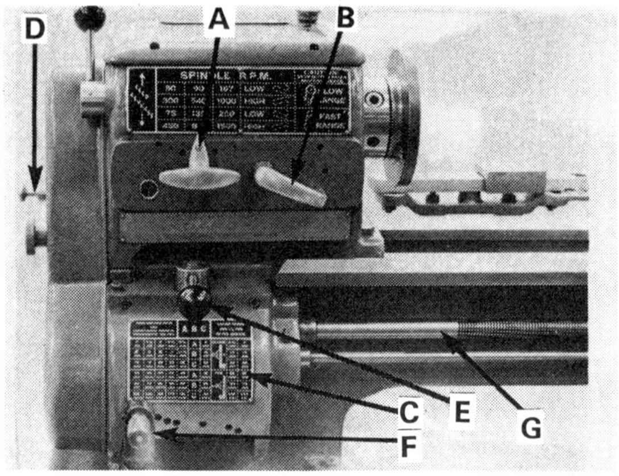

1. **• Set the headstock levers (A and B) so that the lathe spindle revolves approximately one-quarter of the turning speed.**
   • Set the quick-change gearbox (C) to the correct number of threads per inch by setting the levers (D,E,F) to the positions indicated on the chart.
   • Engage the leadscrew (G) if the lathe has a leadscrew engagement lever.

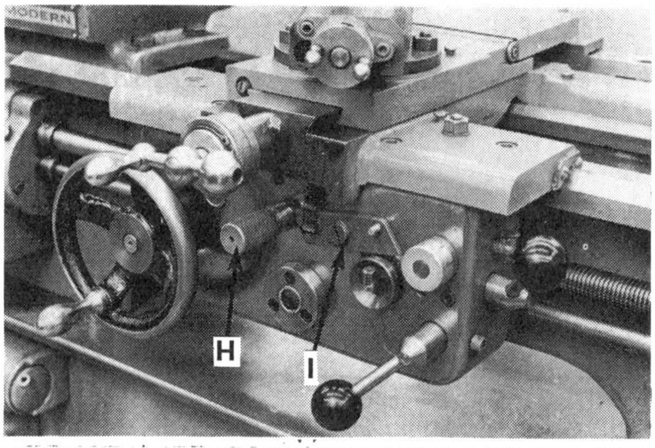

2. **• Set the carriage controls for threading, if the lathe is so equipped.**
   • Place the feed directional plunger (H) in neutral.
   • Engage the locking bar (I) in the neutral groove of the feed directional plunger.

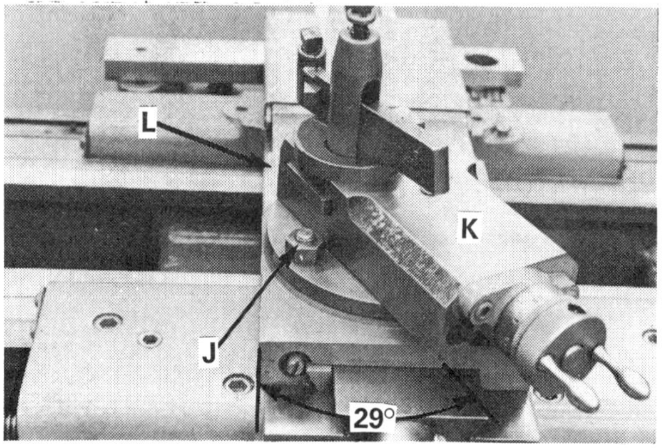

3. • Loosen the compound rest lock nuts (J).
   • Swivel the compound rest (K) 29° to the cross-slide (L).
   • Using two-finger pressure on the wrench, tighten the compound rest lock nuts.

4. • Move the toolpost (M) to the left-hand edge of the compound rest.
   • Mount a 60° threading tool (N) extending approximately ½" (12.7 mm) beyond the toolholder (O).
   • Set the point of the threading tool even with the lathe center point (P).
   • Lightly tighten the toolpost screw (Q).

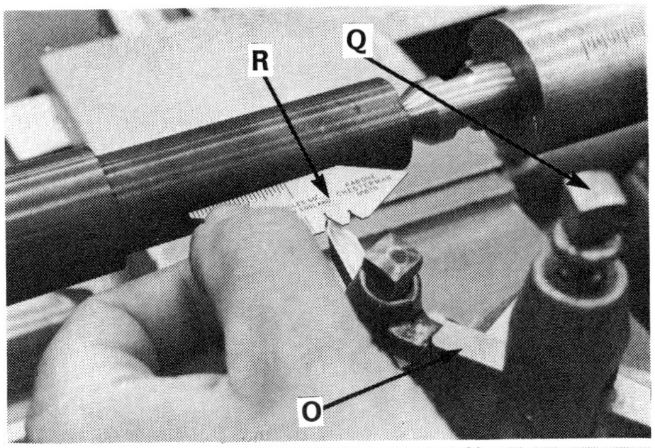

5. ● Mount the work in the lathe.
   ● Hold a thread center gage (R) horizontally on the centerline of the work.
   ● Lightly tap the toolholder (O) until one side of the threading tool aligns exactly with one side of the groove in the gage.
   ● Tighten the toolpost screw (Q) securely.

6. ● With chalk, mark the drive plate slot (S) which is driving the lathe dog.
   ● Measure the length to be threaded with a rule.
   ● At this point, cut a light groove (T) on the work with a threading tool.

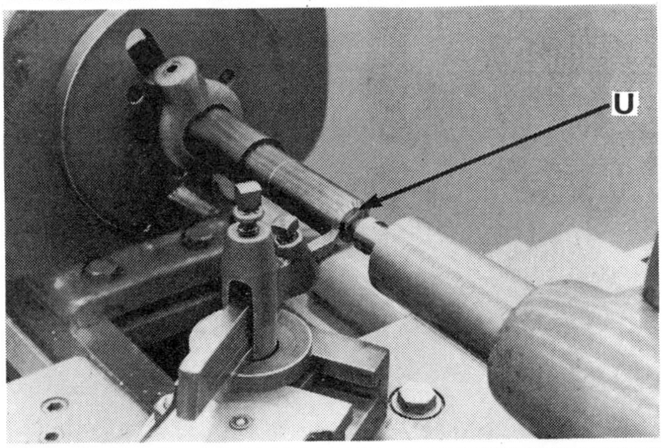

7. ● With the side of the threading tool, cut a chamfer (U) on the end of the work.
   ● The small diameter of the chamfer should be smaller than the minor diameter of the thread to be cut.

8. ● Turn the crossfeed handle (V) to bring the threading tool close to the work diameter and stop when the handle is at the 3 o'clock position.
   ● Hold the crossfeed handle in position with one hand and turn the graduated collar (W) to zero with the other hand.

9. • Move the carriage so that the threading tool is about 1/8" (3.6 mm) from the end of the work.
• Turn the compound rest handle (X) until the tool lightly marks the revolving work.
• Set the compound rest collar (Y) to zero without moving the compound rest handle.
• Move the carriage so that the tool clears the end of the work.

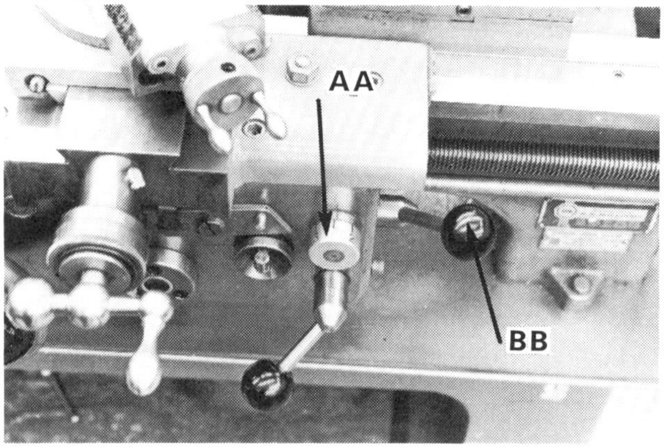

10. • Most thread chasing dials (AA) have four numbered and four unnumbered lines or divisions.
• When cutting an even number of threads per inch, engage the split nut lever (BB) on any line; for odd numbered threads, engage at numbered lines only.

11. • Have the threading tool clearing the end of the work.
• Turn the compound rest handle (X) clockwise .003.
• Engage the split nut lever (BB) on the correct line of the chasing dial (AA) to take a trial cut.

12. • When the threading tool reaches the end of the section to be threaded, withdraw the tool by turning the crossfeed handle (V) *counter-clockwise*.
• Disengage the split nut lever (BB) to stop the carriage movement.

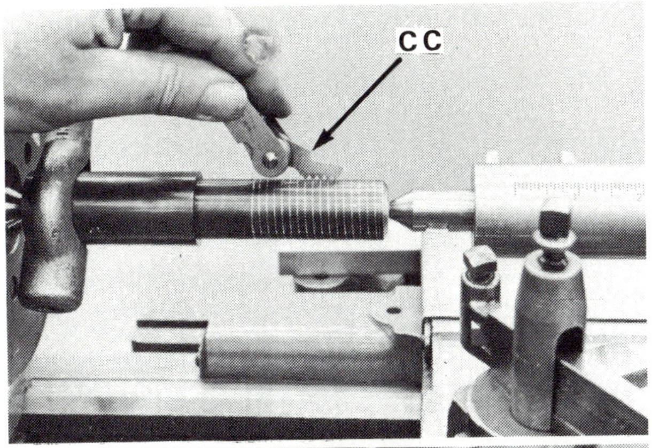

13. • Stop the lathe and move the carriage until the cutting tool clears the end of the work.
    • Use a thread pitch gage (CC) to check the threads per inch of the trial cut.
    • If the trial cut thread does not match the thread pitch gage, recheck the quick-change gearbox setting (Frame #1).

14. • Start the lathe.
    • Return the crossfeed handle (V) to the original zero on the graduated collar (W).
    • Set the depth of each cut using the compound rest handle (X).
    • Engage the split nut lever on the correct line for all successive cuts.

FEED WITH COMPOUND REST AT 29°

T.P.I.	1	2	3	4	5	6	7	8	Finishing Cuts
8	.020	.020	.015	.015	.007	.005	.003	.002	.001
9	.020	.015	.015	.010	.007	.005	.003	.002	.001
10	.020	.015	.015	.010	.004	.003	.002	—	.001
11	.020	.015	.010	.007	.005	.003	.002	—	.001
13	.020	.015	.007	.005	.003	.002	—	—	.001
14	.015	.015	.010	.003	.002	.002	—	—	.001
16	.015	.010	.007	.005	.002	.002	—	—	.001
18	.015	.010	.005	.004	.002	—	—	—	.001

15. • This table should serve as a guide for cutting a good quality thread in the minimum amount of time.
    • In order for these figures to be accurate the toolbit point must be the correct width for each thread pitch.

Width of toolbit point = .125 × pitch of thread.

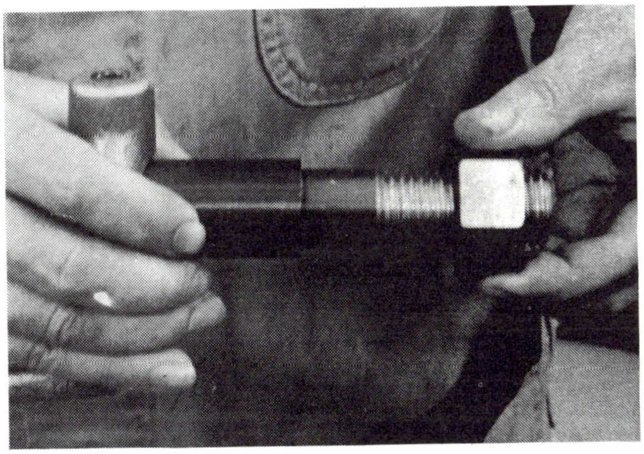

16. • Take successive cuts by feeding the compound rest until the top (crest) and bottom (root) of the thread are the same width.
    • File the burrs from the top of the thread.
    • Test the thread with a master nut.
    • Take finishing cuts until the nut fits properly.

A threading tool must be reset if a partially threaded part has been removed from the lathe or if the threading tool was removed for sharpening. The tool must be accurately reset in the thread groove before the thread can be completed.

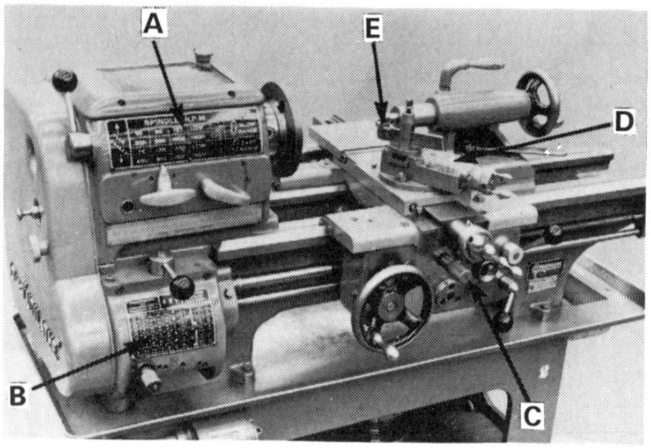

1. • Set the lathe (A) to one-quarter of the turning speed.
   • Set the quick-change gearbox (B) to the correct number of threads per inch.
   • Set the carriage controls (C) for threading.
   • Set the compound rest to 29° (D).
   • Mount and square a threading tool (E) with a thread center gage.

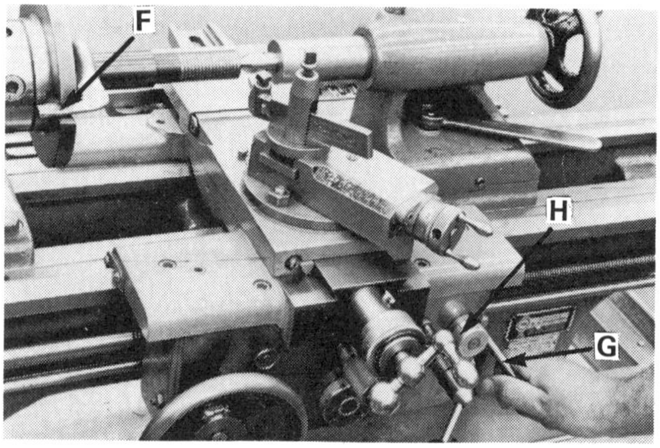

2. • Mount the workpiece.
   • With chalk, mark the drive plate slot (F) that is driving the lathe dog.
   • Start the lathe and engage the split nut lever (G) on the correct line of the thread chasing dial (H).

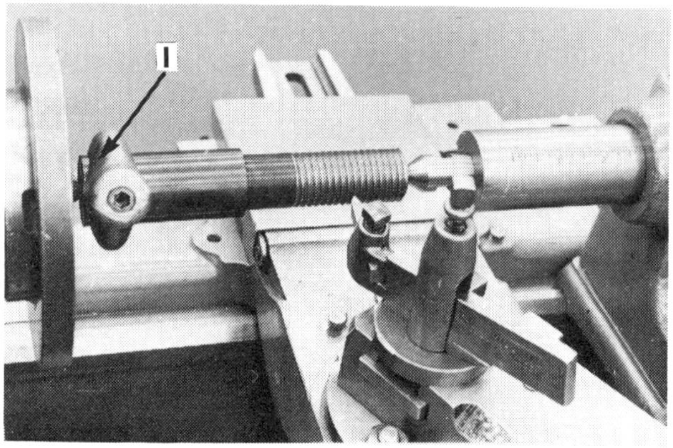

3. • Allow the carriage to travel until the threading tool is over any portion of the unfinished thread.
   • Leave the split nut lever engaged and stop the lathe spindle.
   • Be sure that the lathe dog (I) is in the driving position (against the back) of the drive plate slot.

4. • Leave the split nut lever (G) engaged.
   • By adjusting *only* the crossfeed handle (J) and the compound rest handle (K), fit the threading tool into the partially cut thread.

5. • Adjust *only* the crossfeed and compound rest handles until the side of the threading tool fits against the right hand side of the thread (for right-hand threads).
   • This ensures that no step will remain on the right-hand side when the thread is finished.

6. • Hold the crossfeed handle (J) so that its position does not change and turn the graduated collar (L) to zero.
   • Repeat the same procedure for the graduated collar of the compound rest handle (K).

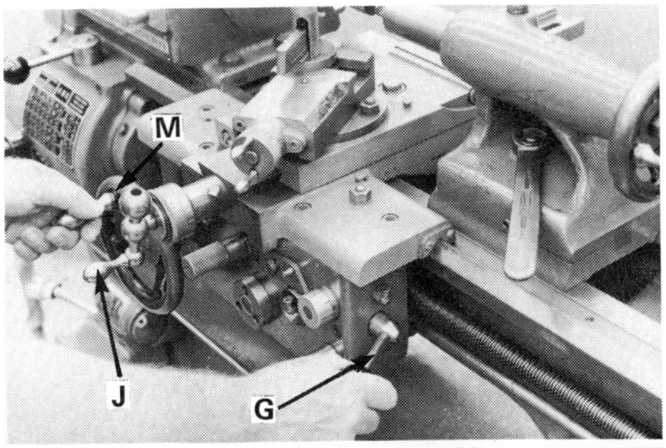

7. • Turn the crossfeed handle (J) counter-clockwise to withdraw the tool from the thread groove.
   • Disengage the split nut lever (G).
   • Turn the carriage handwheel (M) to bring the threading tool to the end of the workpiece.

8. • Start the lathe.
   • Return the crossfeed handle (J) to zero.
   • Set the depth of cut with the compound rest handle (K).
   • Engage the split nut lever (G) on the correct line.
   • Continue threading until the thread is completed.

Lathe accessories such as chucks, faceplates, and drive plates are fitted to the headstock of a lathe. The proper practice for mounting and removing these accessories must be followed in order not to damage the lathe spindle and/or accessories, and to preserve the accuracy of the machine.

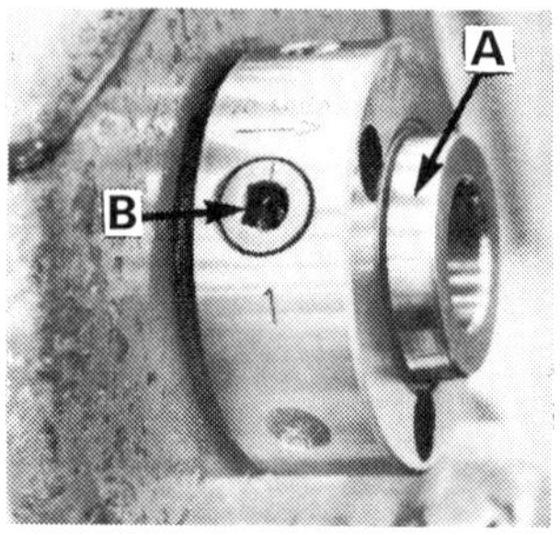

figure A

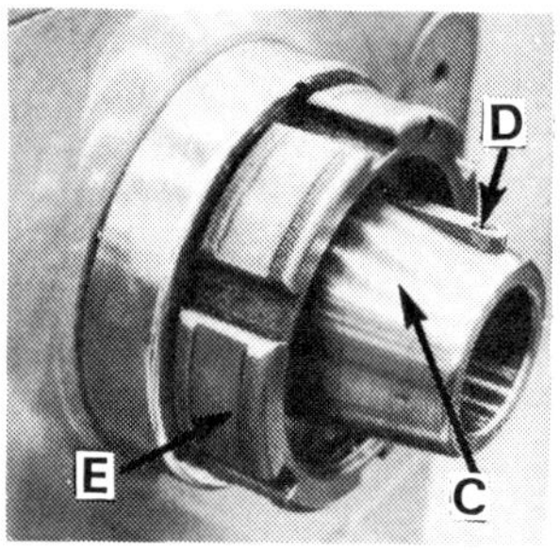

figure B

1.
- *The Cam lock spindle nose* **(figure A) uses a short taper (A) to position the accessory and three or more cam locks (B) to hold the accessory securely on the spindle.**
- *The American standard tapered spindle nose* **(figure B) uses a long steep taper (C) and a key drive (D) to locate the accessory. A lock ring (E) holds the accessory securely on the taper.**

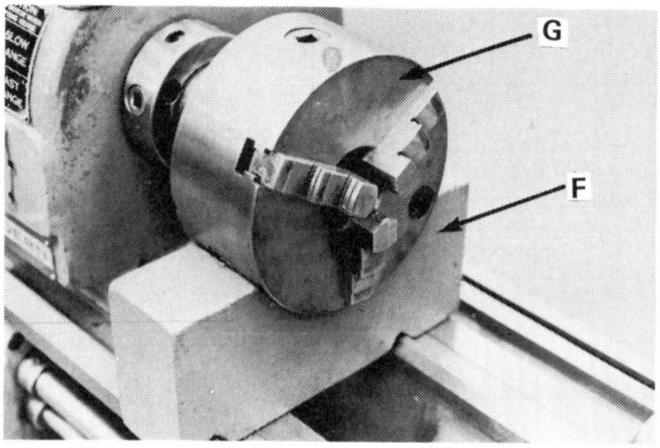

2.
- **Place a suitable chuck block (F) on the lathe bed.**
- **Place the chuck (G) on the chuck block.**
- **The purpose of the chuck block is to prevent injury to the hands or damage to the machine when mounting or removing chucks.**

3. **MOUNTING CAM LOCK CHUCKS**
- **Before mounting the chuck, clean the spindle taper (A) the tapered recess (H) and the cam lock studs (I).**
- **Any dirt or metal chips on the mating surfaces during assembly will produce burrs and impair the accuracy of the machine.**

4.
- **Align the registration lines (J) of the cam locks with the spindle lines.**
- **Align the chuck cam lock studs (I) with the holes in the spindle.**

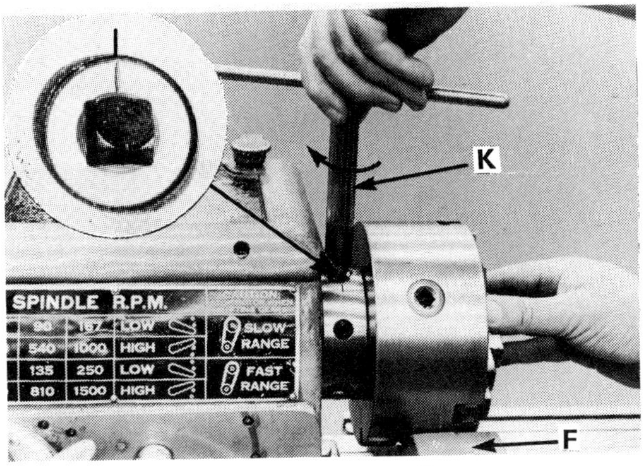

5. • Force the chuck onto the spindle with one hand.
   • With the proper wrench (K) turn each cam lock clockwise until all are tight.
   • Remove the chuck block (F) from the lathe.

6. REMOVING A CAM LOCK CHUCK
   • Place a chuck block (F) under the chuck.
   • Turn the chuck wrench (K) counterclockwise to align the registration line of each cam lock with the lines on the spindle (see insert).

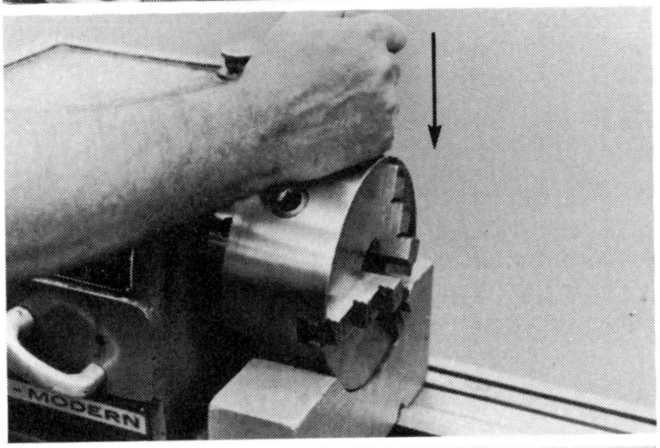

7. • With one hand strike a sharp blow to the chuck to loosen it from the lathe spindle.
   • Remove the chuck and store properly.

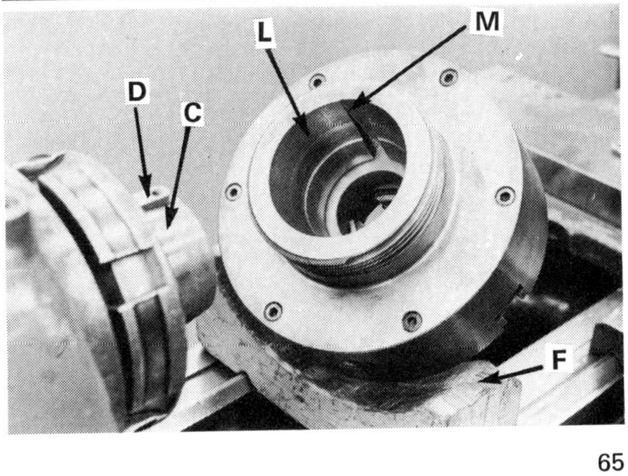

8. MOUNTING A TAPERED SPINDLE NOSE CHUCK
   • Place the chuck on a chuck block (F).
   • Clean the tapered lathe spindle (C) and the tapered hole (L) in the chuck.
   • Align the key (D) in the spindle and the keyway (M) in the chuck.

9. • Slide the chuck onto the spindle being sure that the key and keyway are aligned.
   • Hold the chuck against the spindle with one hand.
   • Tighten the lock ring (E) in a downward direction with the other hand.

10. • Shut off the lathe electrical power.
    • Set the lathe to the lowest spindle speed.
    • Place the proper "C" spanner (N) on top of the lock ring and hold it in position with one hand.
    • With the other hand give the wrench a sharp tap in a *downward* direction to lock the chuck onto the spindle.

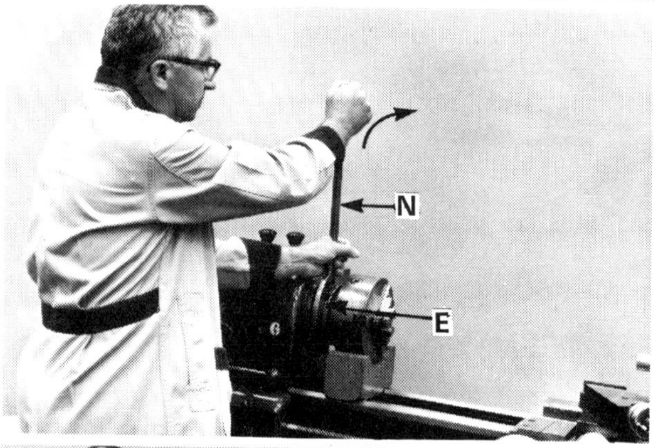

11. REMOVING A TAPERED SPINDLE NOSE CHUCK
    • Place a chuck block under the chuck.
    • Shut off the lathe electrical power.
    • Place the lathe in lowest speed to lock the spindle.
    • Place the wrench (N) on the front of the lock ring (E) with the handle in a vertical position.
    • Hold the wrench on the lock ring with one hand and sharply strike the top of wrench towards the back of the lathe with the other hand.

12. • Loosen the lock ring by hand. (If the lock ring tightens, use the wrench as in Frame 11 to free the chuck from the spindle.)
    • Remove the chuck from the spindle.
    • Plug the tapered hole with a cloth (O) to prevent chips and dirt from entering.
    • Store the chuck in a suitable place.

THREE JAW CHUCK WORK

The *three jaw universal scroll chuck* is used to hold round or hexagonal work for machining. When used properly, this type of chuck is capable of holding work to within .002 of concentricity even after long continued use.

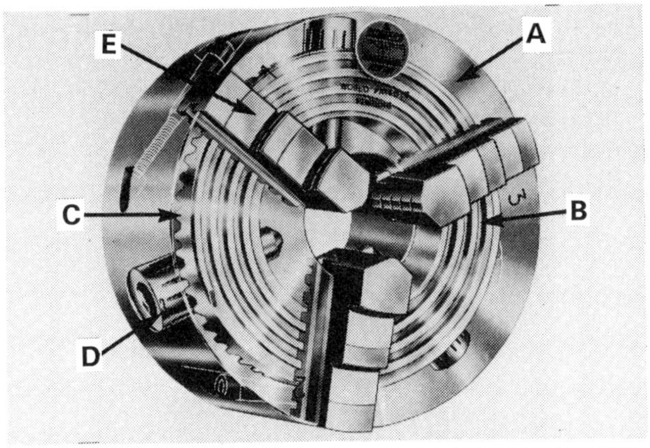

1. A three jaw universal chuck consists of:
   - the body (A) which houses a scroll plate (B)
   - the back of the scroll plate forms a large gear (C) which meshes with three pinions (D)
   - By turning any pinion, the scroll plate turns and advances the three jaws (E) at the same rate since the teeth of the jaws are in mesh with the scroll.

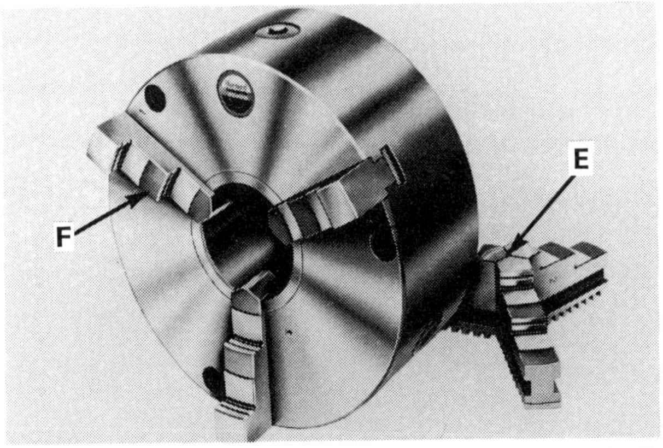

2. Three jaw chucks are supplied with two sets of jaws:
   - a *regular set* (E) for gripping outside diameters and also the inside diameters of larger work.
   - a *reversed set* (F) used to grip the outside of larger diameter work.
   - All jaws are stamped with the same serial number as the chuck in which they are to be used.

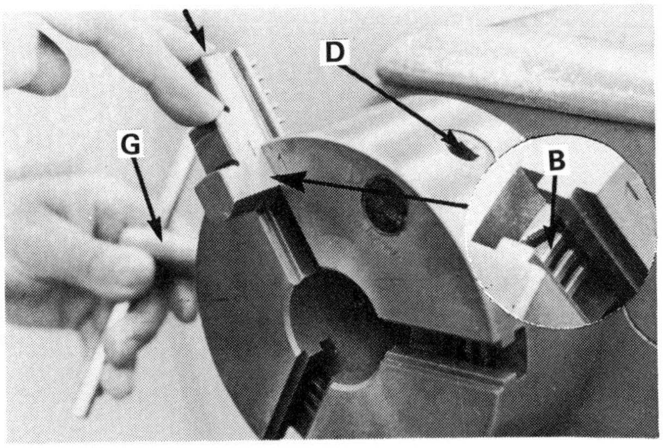

3. ASSEMBLING THE JAWS
   - Clean all jaws and slots.
   - Using a chuck wrench (G) turn one of the pinions (D) until the start of the scroll (B) appears in slot #1 (see insert).
   - Turn chuck wrench counterclockwise one-quarter of a turn.
   - Insert jaw #1 in slot #1.
   - Press down on the jaw while turning the chuck wrench clockwise until the jaw is engaged.

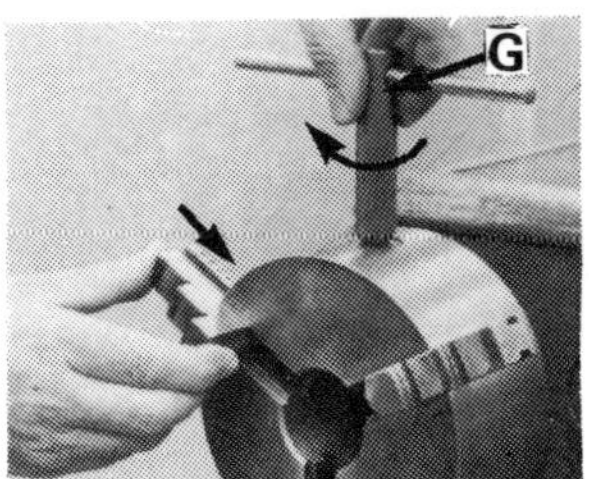

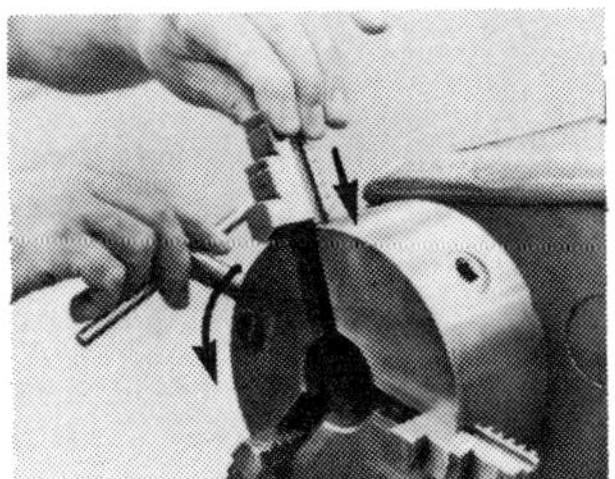

4.
   - Turn the wrench (G) clockwise until the start of the scroll shows in slot #2.
   - Turn the wrench counterclockwise about one-quarter of a turn.
   - Insert jaw #2 into slot #2 and follow the same procedure as for jaw #1.
   - Follow the same procedure for inserting jaw #3.

5. • Open the chuck jaws and insert the workpiece.
   • No more than three times the stock's diameter should extend beyond the chuck jaws. This extension limit is indicated as (X).
   • Rotate round work slightly while tightening the chuck jaws securely.

6. • The outside steps of regular jaws (E) may be used to hold internal diameters.
   • Close the chuck jaws to allow the internal diameter to slip over the jaws.
   • Hold the workpiece firmly against the face of the chuck jaws.
   • Turn the chuck pinion counterclockwise to grip the workpiece.
   • Seat the work against the face of the jaws by tapping with a soft-faced hammer.

7. • Large diameters may be gripped with a set of reversed jaws (F).
   • Assemble the reversed set of jaws as in Frames 3 and 4.
   • Place the workpiece in the chuck and turn the pinion *clockwise* to tighten the jaws on the workpiece.
   • Seat the work against the jaw faces by tapping with a soft-faced hammer.

8. • Set the toolpost (G) to the left of the compound rest (H).
   • Move the carriage until the cutting tool is at the left end of the planned cut.
   • Turn the spindle by hand to see that the chuck jaws do not strike the toolholder or the compound rest.
   • Left-hand toolholders (I) are generally used for turning outside diameters.

DRILLING AND REAMING

Drilling a hole in a workpiece followed by reaming will produce a round, smooth hole of exact size.

1.
- **Mount the work in a suitable chuck.**
- **Set the lathe speed for facing.**

$$\text{rpm} = \frac{\text{CS} \times 4}{\text{Dia. of Work}}$$

- **Face the end to produce a flat smooth surface.**

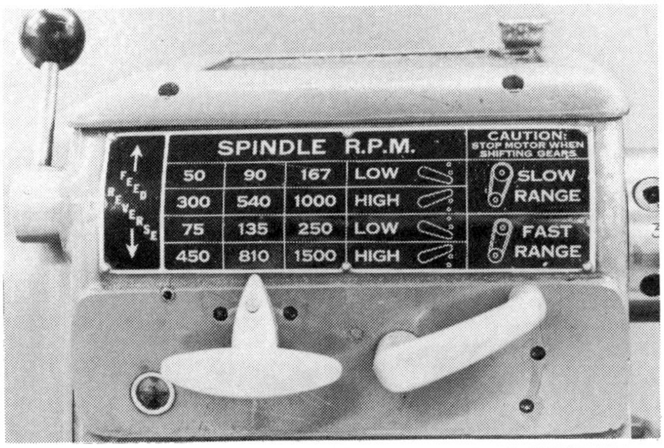

2.
- **Select the proper center drill for the job.**
- **Calculate the proper spindle speed for the center drill being used.**

$$\text{rpm} = \frac{\text{CS} \times 4}{\text{Dia. of center drill}}$$

- **Set the machine to three-quarters of the calculated rpm.**

3.
- **Mount the center drill (A) in the drill chuck (B).**
- **Slide the tailstock (C) up until the center drill nearly touches the work.**
- **Tighten tailstock clamp nut (D).**
- **Start the machine and drill the hole to the proper depth using the tailstock handwheel.**

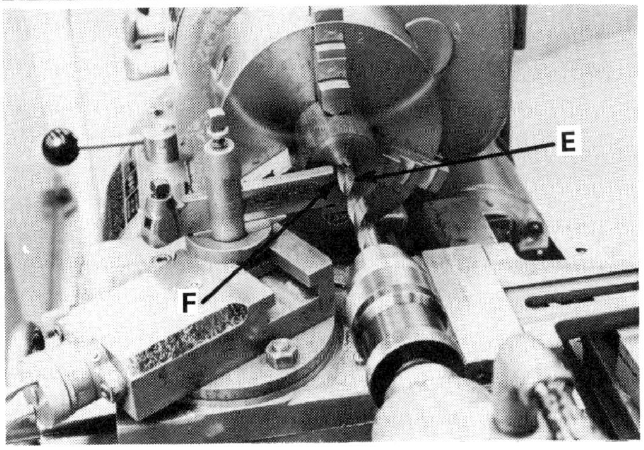

4.
- **Set the proper rpm for the size of the drill you want for the final hole diameter.**
- **Support the diameter of the drill (E) near the point by the back of the toolholder (F) when starting the hole.**
- **Drill the hole to the required depth using the tailstock handwheel.**
- **Use coolant and clear the drill occasionally by reversing its direction with the tailstock handwheel.**

5. USING A TAPER SHANK DRILL
   - Taper shank drills should be prevented from turning by a lathe dog (H).
   - Mount the taper shank drill in the tailstock spindle (G).
   - Place the dog over the drill as close to the tailstock spindle as possible.
   - Place the tail of the dog on the compound rest (I) and tighten the lathe dog set screw.

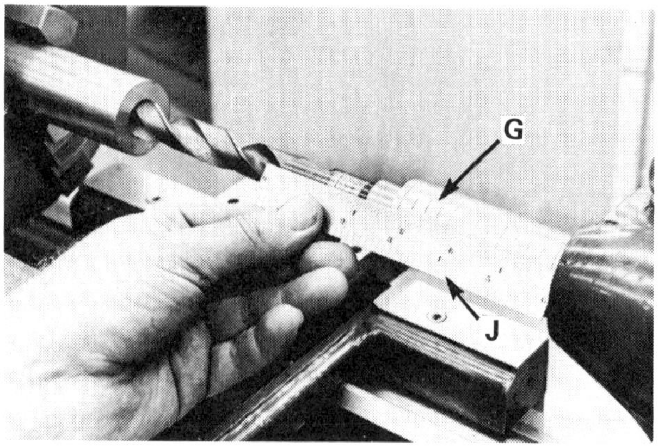

6.
   - Drill the hole to depth using the tailstock spindle graduation (G) or a rule (J) if a more accurate depth is required.
   - The depth of a hole is measured from where the drill just cuts the full diameter.

figure A

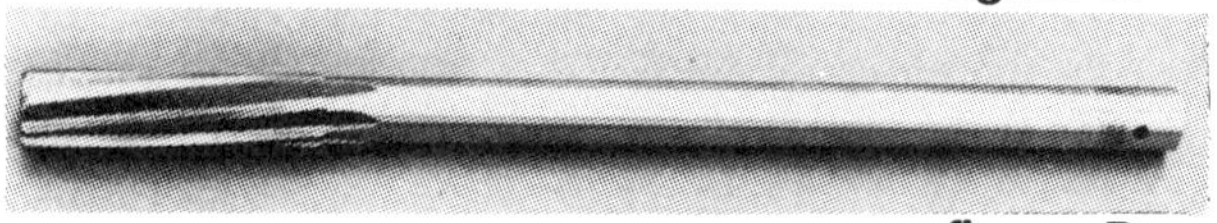

figure B

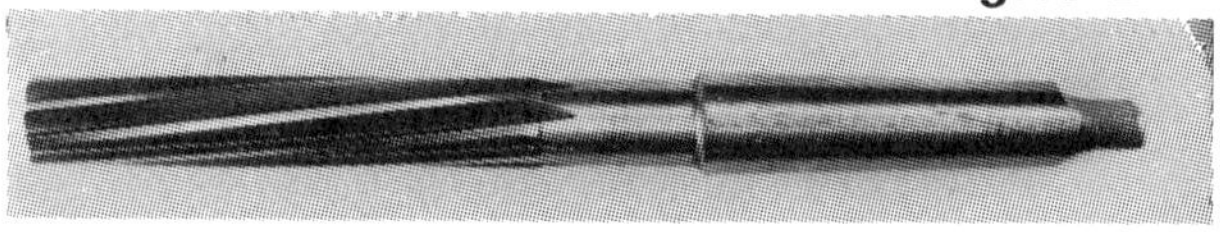

7. REAMING
   - The straight shank machine reamer, figure A, is mounted in a drill chuck, secured in the tailstock.
   - The tapered shank reamer, figure B, is mounted directly in the tailstock spindle. A lathe dog is usually clamped to the upper part of the body to prevent the reamer from turning. (See Frame #5 of this operation sheet.)

8.
   - Mount the proper size reamer .
   - Set the spindle speed to about one-half of drilling speed.
   - Slide the tailstock (C) forward until the reamer almost touches the work, then tighten the tailstock clamp nut (K).
   - Start the machine and apply cutting fluid to the reamer.
   - Feed the reamer into the work using the tailstock handwheel (L).

After a part has been machined, it may be separated from the stock by means of a parting or cutting-off tool. Cutting off is done close to the chuck jaws.

1. • **PARTING OR CUTTING OFF**
   • Grip the workpiece (A) short in the chuck (B).
   • Mount a right-hand offset cutting-off tool (C) in the toolpost (D).
   • Loosen the locking nut (E).
   • Adjust the blade of the tool so that it extends approximately one-half the diameter of the work plus ⅛″ (Y) beyond the toolholder.
   • Tighten the locking nut (E).

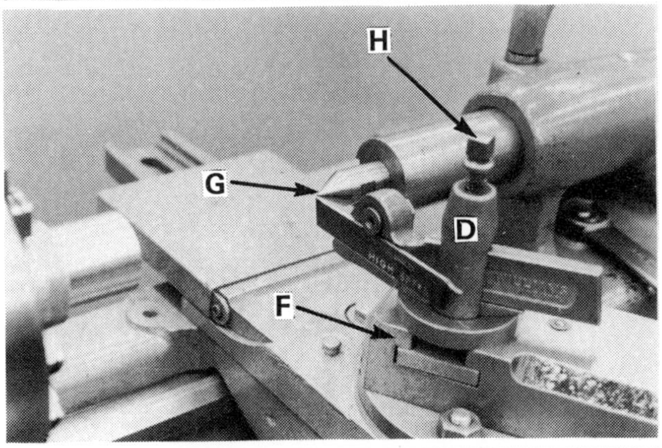

2. • Set the toolpost (D) to the left side of the compound rest (F).
   • Set the cutting tool even with the point of the dead center (G).
   • Lightly tighten the toolpost screw (H).

3. • Move the carriage until the blade is close to the workpiece.
   • Tap the toolholder (C) until the blade is 90° to the workpiece.
   • Tighten the toolpost screw (H) securely.

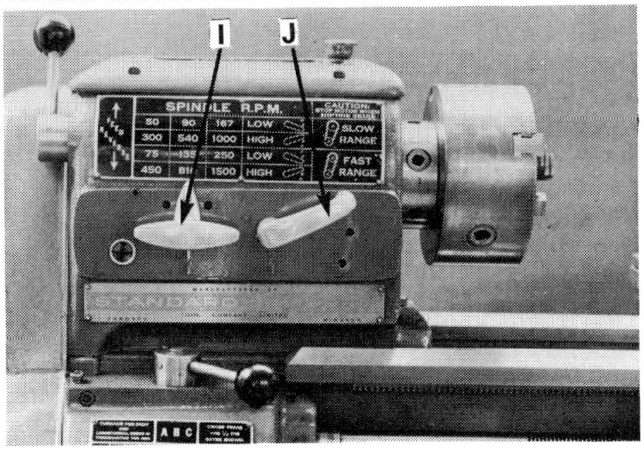

4. • Set the lathe speed to one-half the turning speed by adjusting the headstock levers (I and J) to the proper position.
   • A slower speed will help to reduce the possibility of vibration or chatter during the cutting-off operation.

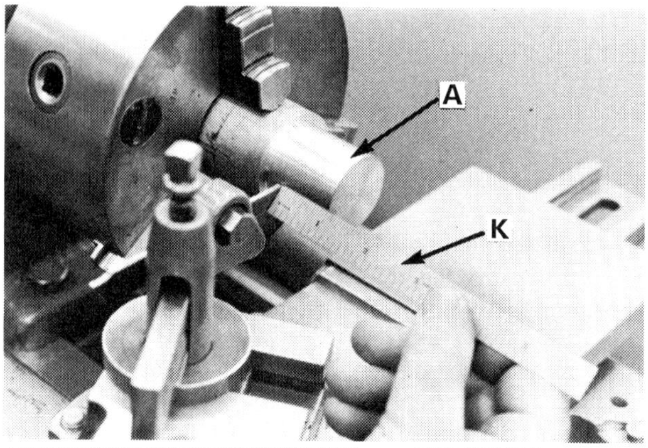

5. • Turn the crossfeed handle counterclockwise untill the tool just clears the work (A).
   • Place a rule (K) against the right side of the cutting tool.
   • Move the carriage until the proper length of the workpiece is indicated on the rule.

6. • Start the lathe.
   • Apply cutting fluid and feed the cutting tool slowly but steadily into the work with the crossfeed handle (L).
   • For work over ½″ in diameter or a groove depth more than twice the width of the cutting tool, move the carriage a few thousandths to the right and left with the carriage handwheel. This prevents the cutting tool from binding.

7. GROOVING
   • Position the cutting tool for length.
   • Apply chalk or layout dye (M) to the work at this point.
   • Start the machine and feed the cutting tool until it just scrapes the chalk or layout dye.
   • Set the crossfeed graduated collar (N) to zero (see insert).
   • Tighten the carriage lock nut. (See Frame 6).
   • Apply cutting fluid and feed the tool to depth using the crossfeed graduated collar.

8. • Only shallow grooves should be cut on work mounted between centers.
   • Work should never be parted off when mounted between centers.
   • Parting off or cutting deep grooves on work mounted between centers will result in bent work.

MOUNTING WORK IN A FOUR JAW CHUCK

When it is necessary for a workpiece to run true, it is generally mounted in a four jaw chuck. Round, square, octagonal and odd-shaped work can be held in this type of chuck.

1. **•** **Measure the diameter of the workpiece.**
   - **Adjust the jaws (A) so they are open slightly larger than the workpiece (B) by using the rings (C) on the face of the chuck (D) to assure that all four jaws are the same distance from the center of the chuck.**
   - **Mount the work and lightly tighten all jaws.**
   - **Adjust all jaws until they are in the same relation to one ring on the chuck face.**

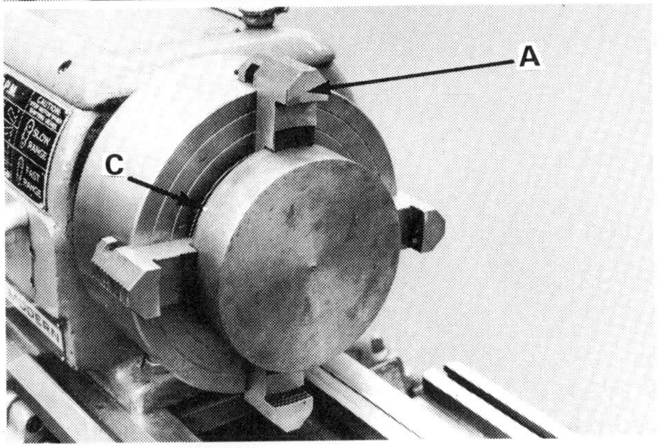

2. **•** **The jaws (A) may be reversed to grip larger work.**
   - **Use the rings (C) on the chuck face to approximately position the work.**
   - **Use any of the following procedures for more accurate method of truing the work.**

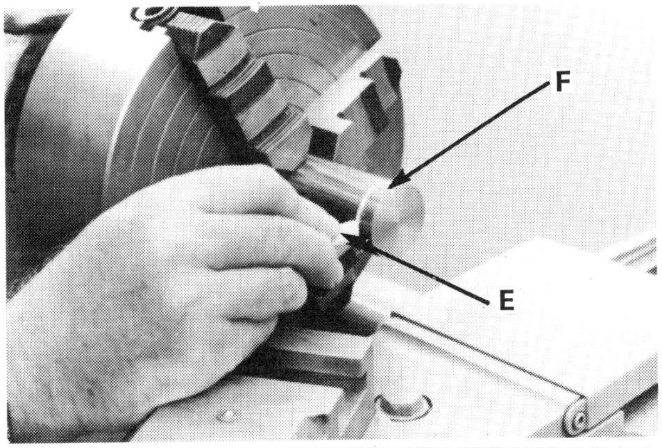

3. **CHALK METHOD**
   - **Start the lathe and slowly bring a piece of chalk (E) against the revolving workpiece.**
   - **Stop the lathe and check the chalk mark.**
   - **Loosen the jaw opposite the chalk mark (F) and tighten the jaw at the chalk mark.**
   - **Repeat this procedure until the chalk mark is uniform around the circumference.**

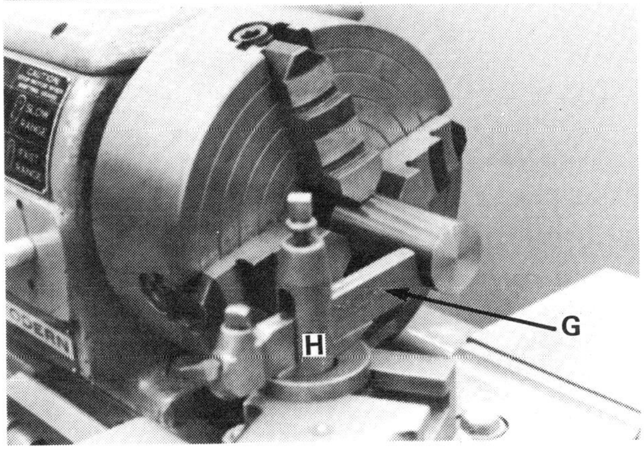

4. **TOOLHOLDER METHOD**
   - **Reverse the toolholder (G) in the toolpost (H).**
   - **Bring the back of the toolholder close to the workpiece.**
   - **Revolve the chuck by hand and note the high and low spots.**
   - **Loosen the jaw at the low spot and tighten the opposite jaw.**
   - **Repeat this procedure until the work runs true.**

5. SURFACE GAGE METHOD
   - Place a surface gage (I) on the carriage and bring the scriber (J) close to the workpiece.
   - Revolve the chuck by hand to locate the high and low spots.
   - Loosen the jaw at the low spot and tighten the opposite jaw.
   - Repeat this procedure until the work runs true.

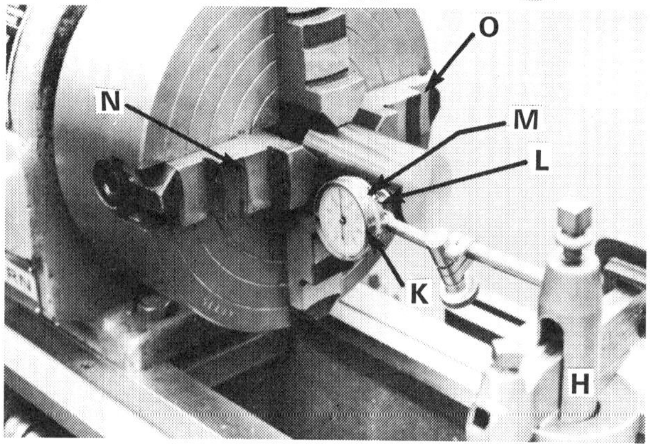

6. DIAL INDICATOR METHOD
   - True the work by the chalk, surface gage, or toolholder method.
   - Mount an indicator (K) in the toolpost (H).
   - Set the indicator plunger (L) to center and in a horizontal position.
   - Adjust the crossfeed handle until the indicator needle registers .030.
   - Revolve the chuck by hand to locate the high spot and set the indicator bezel (M) to zero.

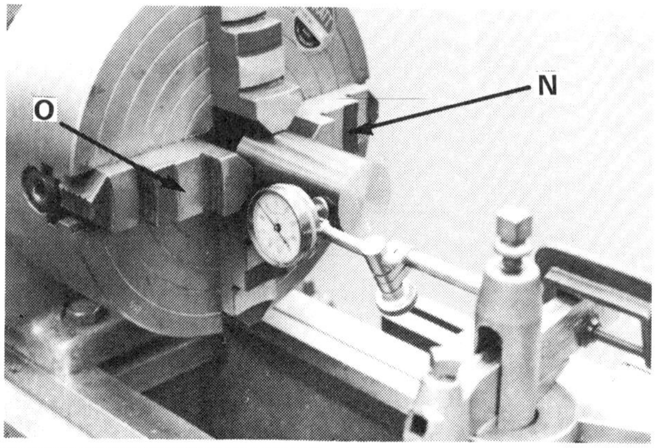

7.
   - Revolve the chuck one-half turn and note the reading at the low spot.
   - Loosen the jaw (O) at the low reading slightly and tighten the opposite jaw (N).
   - Move the work one-half the difference of the indicator readings.
   - Adjust these two jaws only until the indicator reading is the same for each jaw.
   - Repeat this procedure for the other two jaws.

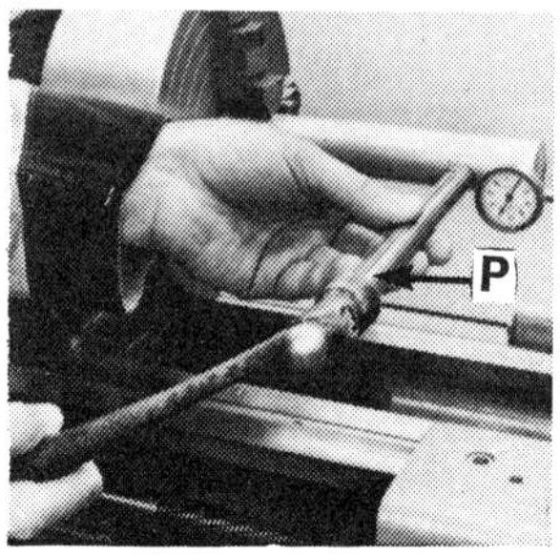

8. PROCEDURE FOR LONG WORK
   - Tighten all jaws and recheck the indicator reading.
   - Follow Frames 6 and 7 to align the work at the headstock.
   - Bring the indicator near the end of the work.
   - Revolve the chuck to locate the high and low readings.
   - With a soft rod, (P) tap the high side away from the indicator until the needle moves half the difference.
   - Repeat procedure until the work runs true at both ends.

Most work machined in a shaper is held in a vise and supported on parallels to raise it to the proper height. In order to produce accurate work, burrs must be removed from the work and the vise and parallels must be thoroughly cleaned.

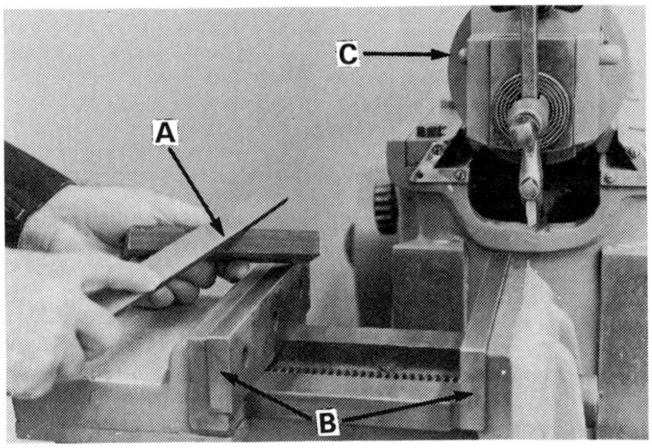

1. • **Remove the burrs from the work with a flat file (A).**
   • **Remove all burrs from both vise jaws (B) with a honing stone.**
   • **Clean the work and the vise.**
   • **Align the fixed jaw of the vise so that it is parallel to the travel of the shaper ram (C).**

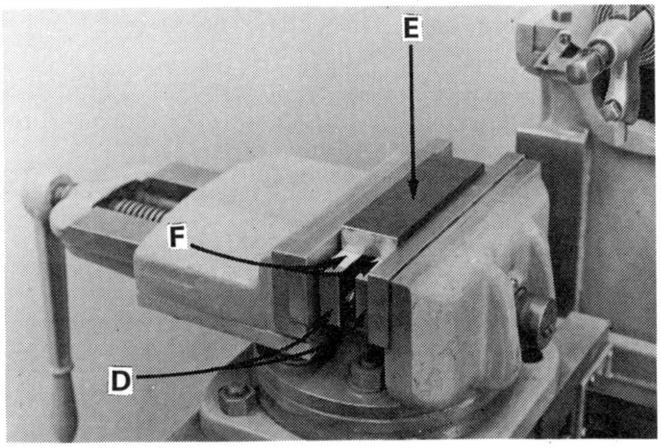

2. • **Move the vise clear of the toolhead.**
   • **Select parallels (D) large enough to hold the work about ¼" (6.35 mm) above the vise jaws.**
   • **Set the work (E) on the parallels in the center of the vise.**
   • **Place a short paper feeler (F) between each corner of the work and the parallels.**

3. • **Tighten the vise securely.**
   • **If the work should raise up in the vise due to rounded or uneven edges:**
   (1) Remove the work from the vise.
   (2) Fold a piece of abrasive cloth, about the length of the workpiece so that both abrasive edges are facing out.
   (3) Place the abrasive cloth between the work and vise jaw near the bottom of the edge which has been raised up.
   (4) Retighten the vise securely.

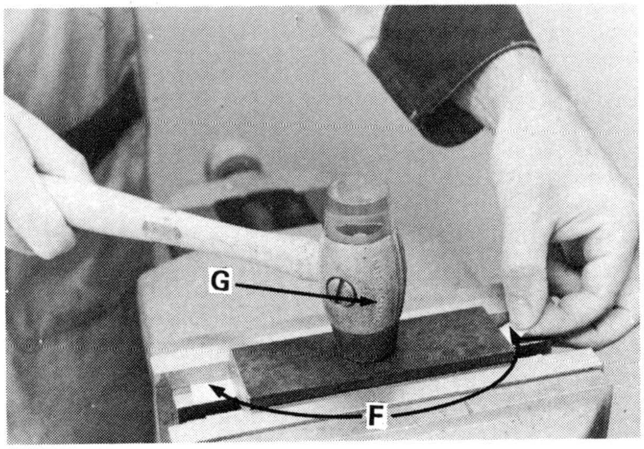

4. • **Lightly tap the work down onto the parallels with a soft-faced hammer (G).**
   • **Alternate hammer blows between both ends of the work.**
   • **Check that all paper feelers (F) are tight.**

SETTING THE LENGTH AND POSITION OF THE SHAPER STROKE

To efficiently machine work in a shaper, the length of the shaper stroke should be set at about an inch more than the length of the workpiece. This will reduce the waste time caused by overtravel of the ram.

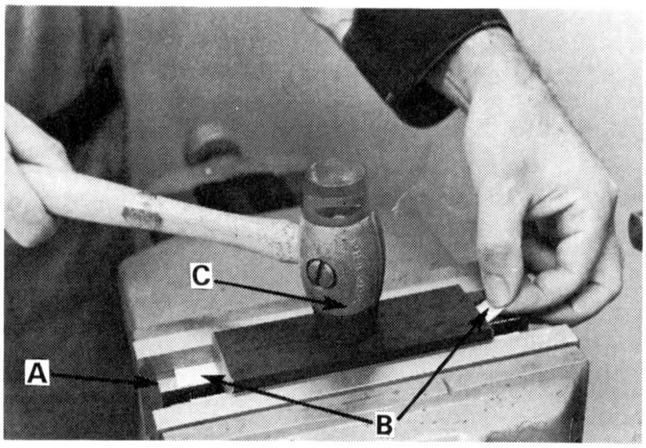

1. • Remove all burrs and clean the work and vise.
   • Set the work on parallels (A) in the center of the vise, about ¼″ (6.35 mm) above the vise jaws.
   • Place short paper feelers (B) between each corner of the work and the parallels.
   • Tighten the vise securely.
   • Tap the work down with a soft-faced hammer (C) until the paper feelers are tight.

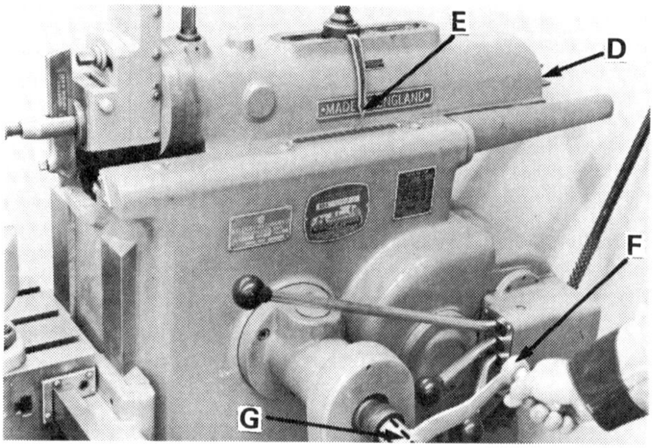

2. • Move the work clear of the cutting tool.
   • Set the shaper to the lowest speed.
   • Use the start-stop or inching button to jog the motor until the ram (D) is at the extreme back of the stroke, indicated by pointer (E).

   OR

   • If the shaper is not equipped with an inching button:
     (1) Place the crank (F) on the stroke regulator shaft (G).
     (2) Turn the crank until the ram reaches the back of the stroke.

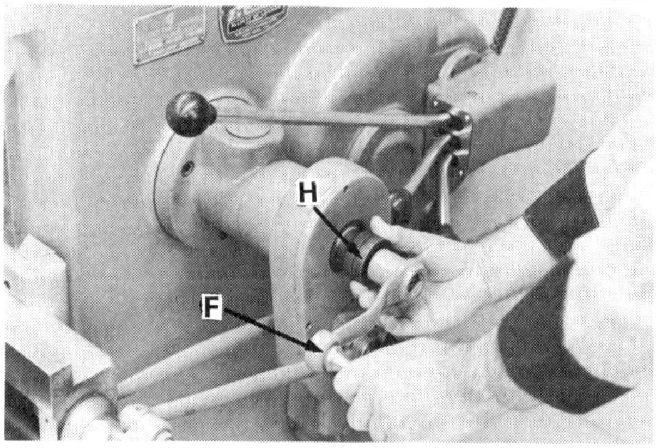

3. • Stop the motor.
   • Grip the stroke regulator shaft lock-nut (H) firmly by hand.
   • With the other hand, turn crank (F) clockwise to unlock the stroke regulator shaft nut.
   • Loosen the lock-nut about one-half a turn.

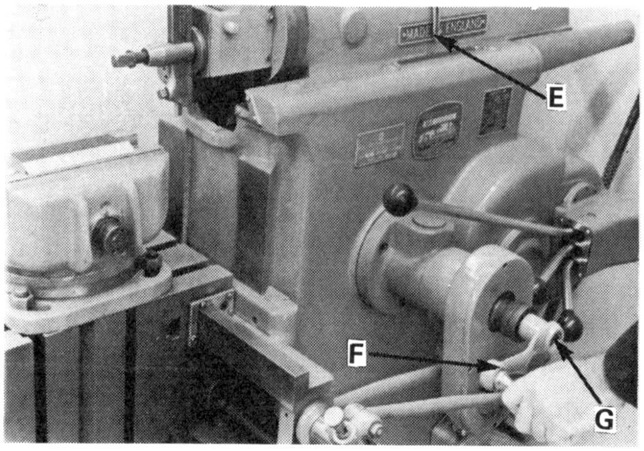

4. • Measure the length of the work and add about 1″ (25.4 mm) for tool clearance (½″ or 12.7 mm at each end of the stroke).
   • Turn the stroke regulator shaft (G) with the crank (F) until the pointer (E) reaches the desired length of stroke.

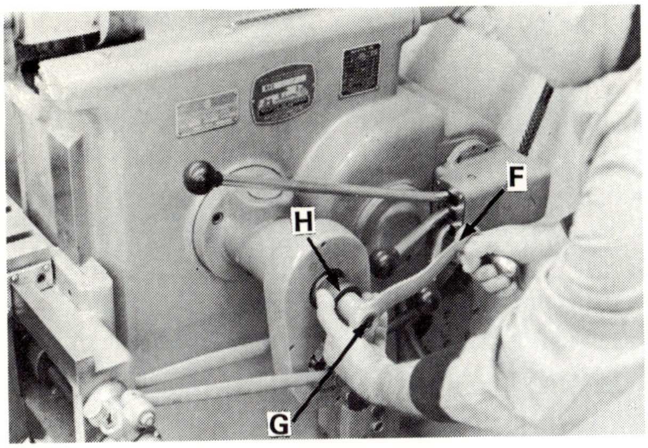

5. • Tighten the lock-nut (H) by hand.
   • Hold the lock-nut firmly and turn the crank (F) counterclockwise to lock the stroke regulator shaft (G).
   • *Remove the crank* from the stroke regulator shaft.

6. • Loosen the ram lock (I) with the crank (F). (This will allow the ram to be positioned in relation to the workpiece.)
   • DO NOT start the machine when the ram lock is loose.

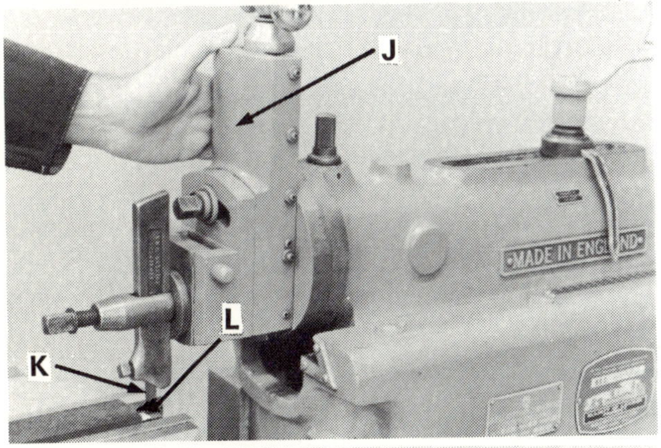

7. • Pull the toolhead (J) forward until the toolbit (K) is within ½″ (12.7 mm) of the work (L).

   OR

   • If the ram is equipped with an adjusting screw, turn the adjusting screw until the toolbit is within ½″ (12.7 mm) of the work.

8. • Tighten the ram lock (I).
   • With the cutting tool (K) clear of the work:
     (1) Start the machine.
     (2) Check that the toolbit clears each end of the work by ½″ (12.7 mm).

   OR

   • With the machine shut off:
     (1) Turn the crank on the stroke regulator shaft until the ram makes one complete stroke.
     (2) Check that the toolbit clears each end of the work by ½″ (12.7 mm).

Most work machined in a shaper is held in a vise. It may be held directly by the vise jaws or by using hold-downs between the vise jaws and the workpiece.

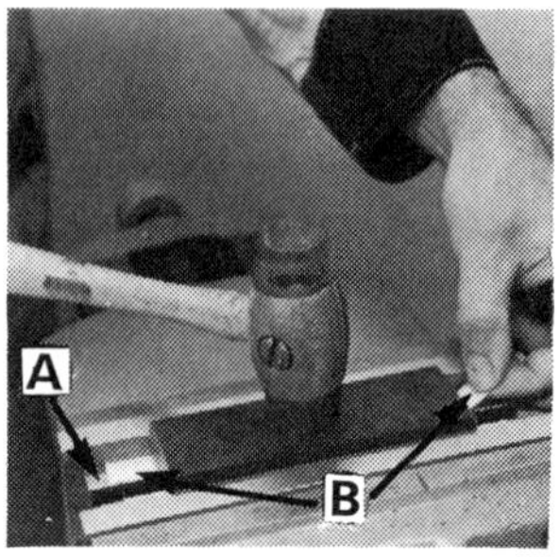

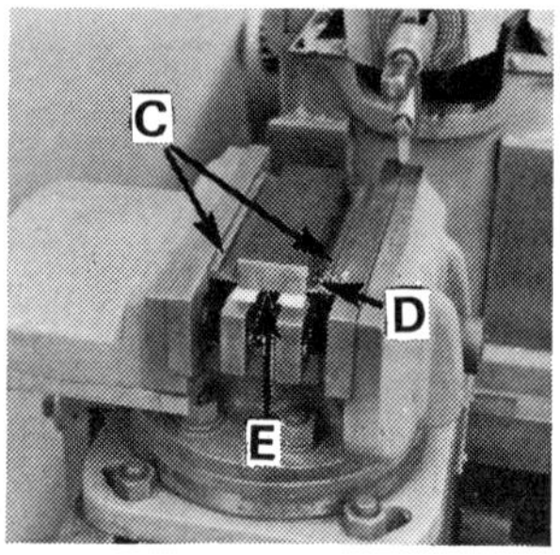

1. - **Remove all burrs and clean the work and vise.**
   - **Set the work on parallels (A) and paper feelers (B).**
   - **Tighten the vise securely and tap the work down until all paper feelers (B) are tight.**

   To Hold Work using Hold-downs (C):
   - Place both hold-downs (C) between each vise jaw and the work with its thin edge (D) close to the bottom of the work (E).
   - **Be sure that both hold-downs are at the same height and tighten the vise securely.**

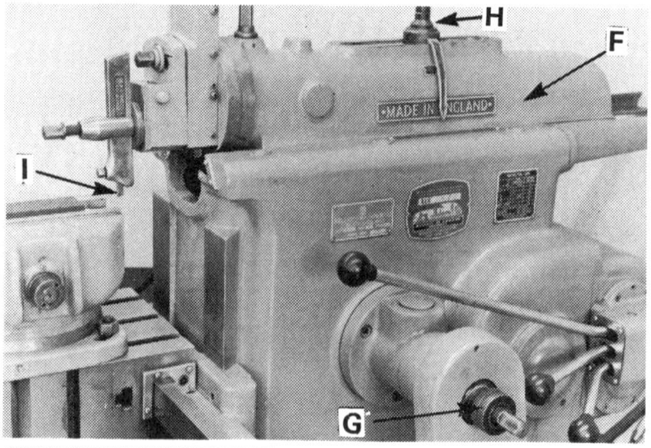

2. - Stop the shaper when the ram (F) is at the back end of the stroke.
   - Loosen the stroke regulator lock-nut (G) and set the shaper stroke to the correct length.
   - Tighten the stroke regulator lock-nut.
   - Loosen ram lock (H).
   - Move the ram (F) to position the toolbit (I) within ½" (12.7 mm) of the work.
   - Tighten the ram lock.

3. - Calculate the shaper strokes per minute (N) using: $N = \frac{\text{Cutting speed} \times 7}{\text{Length of stroke}}$
   - Stop the shaper.
   - Check the speed chart (J) and locate the speed closest to the calculated speed.
   - Set the levers (K and L) to the positions indicated on the speed chart (J).

4. - The feed on a shaper operates on the return stroke of the ram. The clapper box (M) should be set to allow the cutting tool to raise on the return stroke and not rub against the work and become dull too quickly.
   - Loosen the apron lock (O).
   - Swing the top of the apron in the direction of the arrow.
   - Tighten the apron lock and set the toolholder (P) vertically.

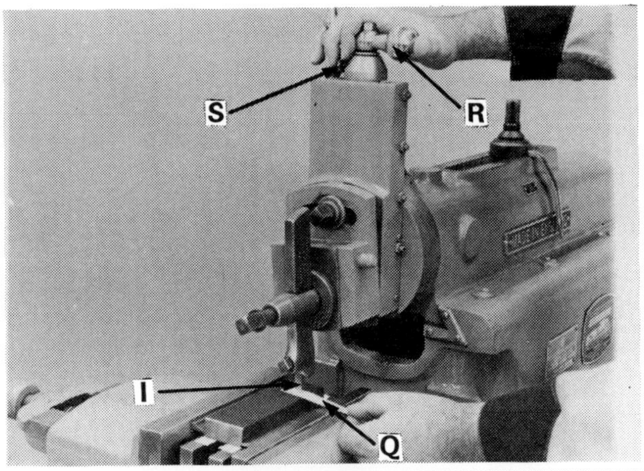

5. To set the Toolbit to the Work:
   - Start the shaper and stop it when the cutting tool (I) is over the work.
   - Hold a paper feeler (Q) under the cutting tool.
   - Feed the toolhead down by turning the downfeed handle (R) until a slight drag is felt on the paper.
   - Loosen the lock screw (S) and set the graduated collar to zero.
   - Tighten the graduated collar lock screw (S).

6. Before setting the depth of cut, the table must be moved so that the cutting tool (I) clears the edge of the work (T).
   - Place the crank (U) on the crossfeed traverse screw (V).
   - Turn the crank (U) clockwise to move the table and work so that the toolbit clears the edge of the work by about ¼" (6.35 mm).

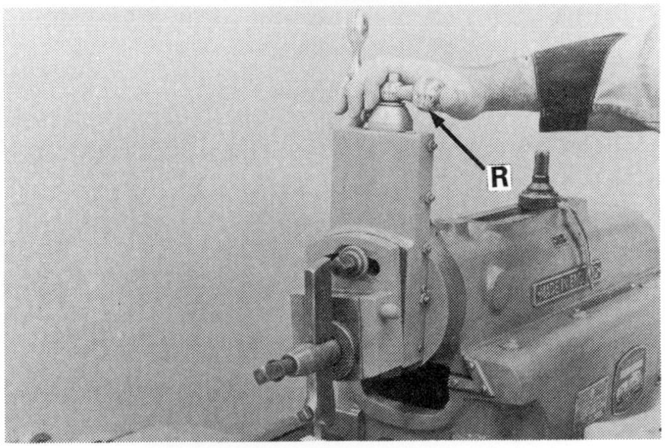

7. About .060 (1.52 mm) should be removed from a surface in order to produce a flat surface.
   - Set the depth of cut by turning the downfeed handle (R) clockwise the required amount.

8. The shaper table feed should be .005 to .015 (.12 to .38 mm) per stroke.
   - Loosen the connecting rod lock (W).
   - Adjust the connecting rod (up or down) until the correct feed is set.
   - Tighten the connecting rod lock.
   - Start the shaper and see that the feed operates on the return stroke of the ram.
   - Engage the pawl (X) and machine the surface.

In order to machine the four sides of a piece of work square and parallel, it is important that each side be machined in a definite order. It is very important that dirt and burrs be removed from the work, vise, and parallels, since they can cause inaccurate work.

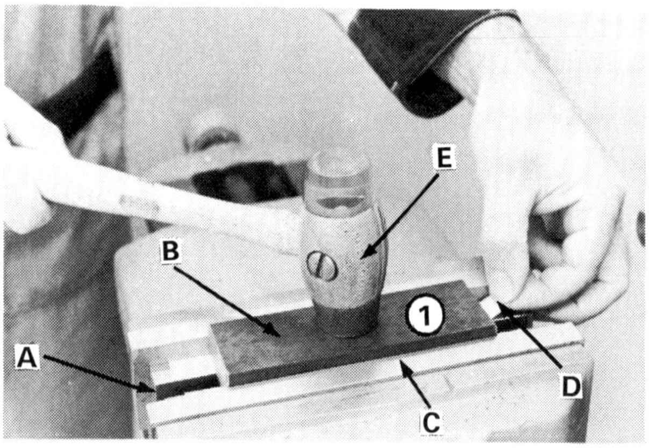

1. SIDE #1
   - Remove all burrs from the work, parallels and vise and clean them thoroughly.
   - Select parallels (A) which will raise the work (B) ¼" (6.35 mm) above the vise jaws (C).
   - Set the work, with its largest side (#1) up, in the center of the vise.
   - Place a short paper feeler (D) between each corner of the work and the parallels.
   - Tighten the vise securely.
   - Lightly tap the work down with a soft-faced hammer (E) until all paper feelers (D) are tight.

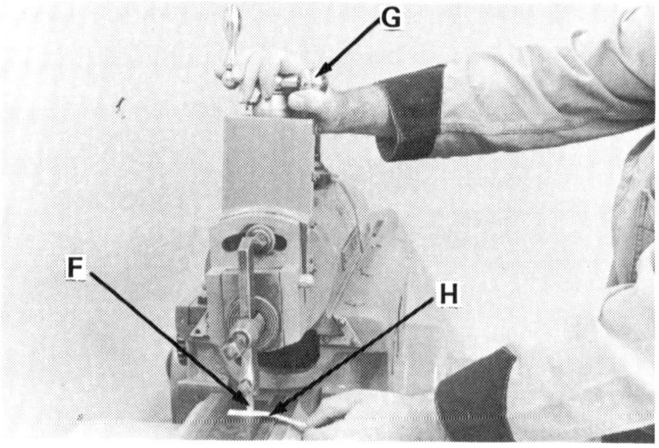

2. - Set the length and position of the stroke (Operation #2).
   - Move the ram forward until the cutting tool (F) is over the work.
   - Turn the downfeed handle (G) until the toolbit (F) is close to the work.
   - Place a paper feeler (H) between the cutting tool and the work.
   - Lower the toolhead until a slight drag is felt on the paper between the work and the toolbit.

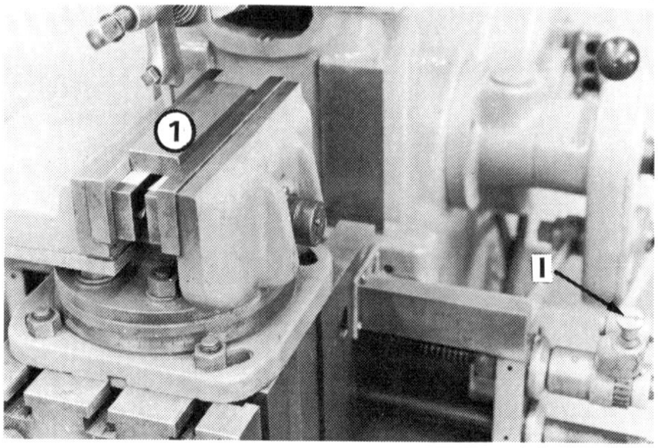

3. - Move the table so that the cutting tool clears the right-hand edge of the work.
   - Set the shaper speed and feed.
   - Lower the cutting tool about .040 (1.01 mm) using the graduated collar on the downfeed handle.
   - Start the shaper.
   - Move the table with the crossfeed traverse crank until the work contacts the cutting tool.
   - Engage the automatic feed (I) and machine side #1.

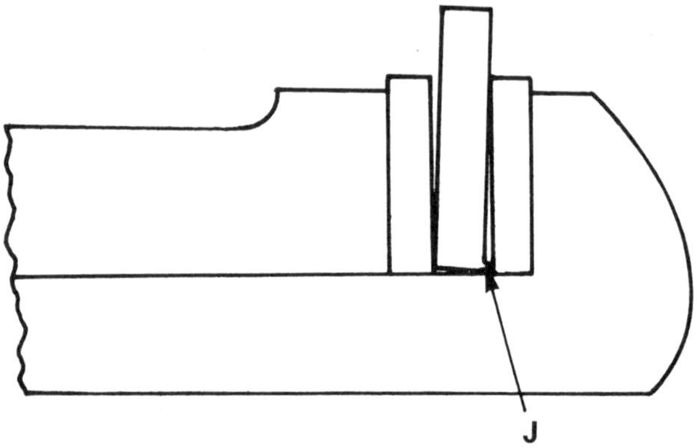

4. - Remove the work from the vise.
   - File off *all burrs* from the edges of the surface just machined.
   - If the burrs are not removed such as at (J) they will stop the machined surface from being held flat against the solid vise jaw.
   - If the work is tilted and not held flat against the solid jaw of the vise and side #2 is machined, it will not be square (at 90°) with side #1.

5. MACHINING SIDE #2
   - Clean the vise and work thoroughly.
   - Set the work on parallels, if necessary, with side #1 against the solid (fixed) jaw (K) and side #2 up.
   - Place a round bar (L) between side #4 and the movable vise jaw (M).
     NOTE: Be sure that the round bar is in the *center of the amount of work being held inside the vise jaws.*
   - Tighten the vise securely and tap the work down until the paper feelers (D) are tight.

6.
   - Set the cutting tool to the work surface using a paper feeler.
   - Move the table so that the cutting tool clears the right-hand edge of the work.
   - Lower the cutting tool about .040 (1.01 mm) using the downfeed graduated collar.
   - Start the shaper.
   - Move the table until the work contacts the cutting tool.
   - Engage the automatic feed and machine side #2.
   - Remove the work, file off all burrs and check that side #1 and #2 are square with each other.

7. MACHINING SIDE #3
   - Place side #1 against the solid jaw (K) with side #2 resting on the parallels.
   - Place a round bar (L) between side #4 and the movable jaw.
     NOTE: The round bar *must* be in the center of the amount of work being held *inside* the vise jaws.
   - Tighten the vise securely and seat the work on the parallels.
   - Set the proper depth of cut and machine side #3 to size.
   - Remove the work and file off all burrs.

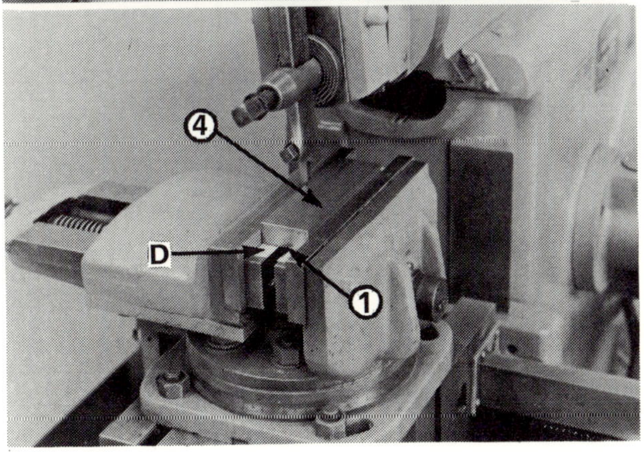

8. MACHINING SIDE #4
   - Set side #1 on parallels using paper feelers (D) under each corner of the work.
   - Tighten the vise securely and tap the work down until all feelers are tight.
   - Set the cutting tool for the proper depth of cut.
   - Start the shaper and engage the feed.
   - Machine side #4 to size.
   - Remove the work and file off all burrs.
   - Check the work for size and squareness.

When it is necessary to machine work parallel or at right angles to the travel of the shaper ram, the vise jaw must be aligned. The two items commonly used to align the vise are a master square and a dial indicator.

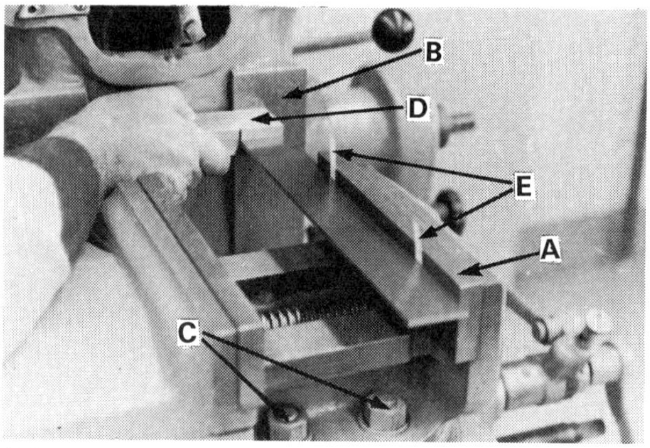

1. **ALIGNING WITH A SQUARE**
   - Clean and remove all burrs from the solid vise jaw (A) with a honing stone.
   - Clean the surfaces of the column (B).
   - Loosen each swivel base lock nut (C) about one-half of a turn.
   - Hold the square (D) against the column (B).
   - Place two paper feelers (E) between the blade of the square (D) and the solid jaw (A).
   - Tap the vise with a soft-faced hammer until both paper feelers are tight.
   - Tighten the swivel base lock nuts.

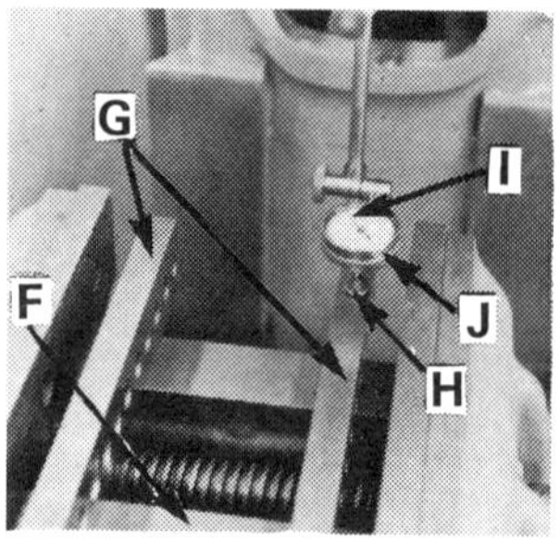

2. **CHECKING THE VISE SEAT FOR PARALLELISM**
   - Remove all dirt and burrs from the vise base (F).
   - Place a set of parallels (G) on the vise base.
   - Mount a dial indicator assembly in the toolpost.
   - Lower the toolhead until the indicator plunger (H) contacts the top of the parallel and the needle (I) registers about ¼ of a turn.
   - Set the indicator bezel (J) to zero.
   - Move the ram by hand until the indicator travels the length of the parallel (G).
   - Note the indicator reading at each end of the parallel.

3.
   - Move the table until the indicator plunger (H) bears on the other parallel (G).
   - Move the ram by hand until the indicator travels the full length of the parallel.
   - Note the reading on the indicator. Any movement of the needle on either parallel indicates that the base of the vise is not parallel.
   - Remove the vise from the machine.
   - Remove all burrs from the vise base and table.
   - Re-assemble and recheck the vise.

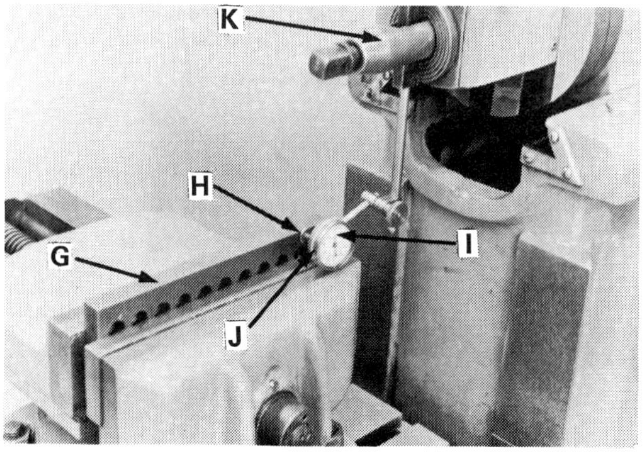

4. **ALIGNING THE VISE PARALLEL TO THE RAM TRAVEL**
   - Tighten a suitable parallel (G) in the vise with about 1" (25.4 mm) projecting above the vise jaws.
   - Mount a dial indicator assembly in the toolpost (K).
   - Lower the toolhead and move the table until the indicator plunger (H) contacts the side of the parallel.
   - Adjust the table until the indicator needle (I) registers about one-quarter of a turn.
   - Set the bezel (J) to zero.

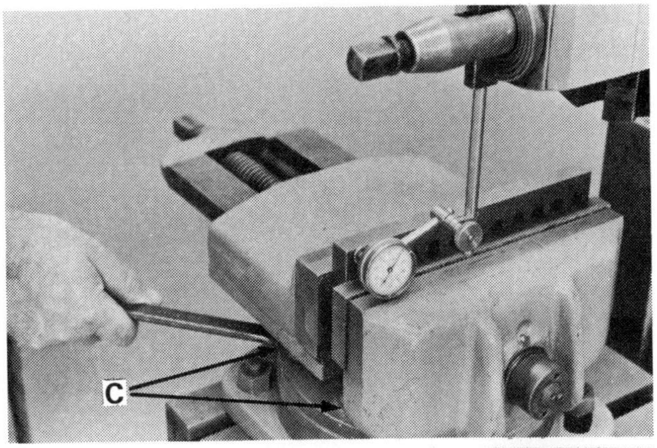

5. • Move the ram by hand until the indicator is at the other end of the parallel.
   • Note the reading on the dial indicator. (Any difference in the indicator reading at each end of the parallel means that the vise jaw is not parallel to the travel of the ram.)
   • Loosen each swivel base lock nut (C) about one-half of a turn.

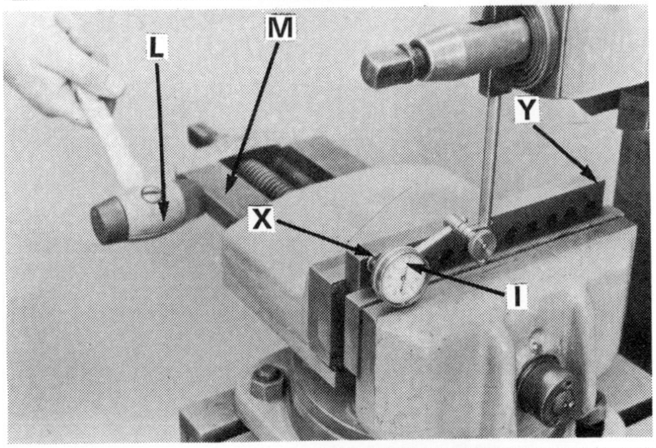

6. • Note where the indicator reading (I) is higher.
   (a) If the reading is higher at X, use a soft-faced hammer (L) to tap the vise (M) clockwise one-half the difference in the indicator readings.
   (b) If the reading is higher at Y, tap the vise counterclockwise.

   NOTE: Always tap the vise so that the parallel moves away from and not towards the indicator.
   • Tighten the swivel base lock nuts and recheck the alignment.

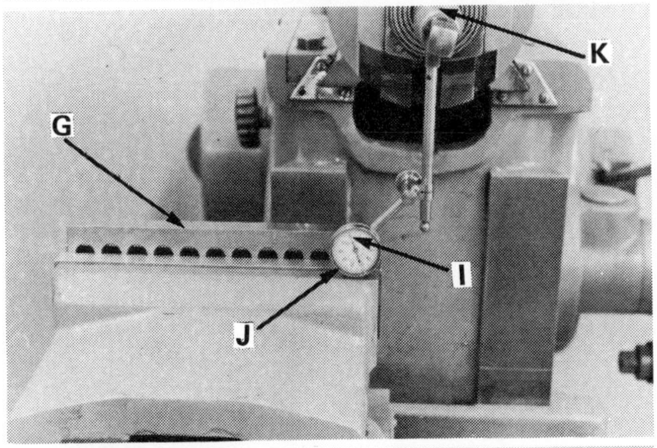

7. ALIGNING THE VISE AT 90° TO THE RAM TRAVEL
   • Tighten a suitable parallel (G) in the vise with about 1" (25.4 mm) projecting above the vise jaws.
   • Mount a dial indicator assembly in the toolpost (K).
   • Adjust the table and move the ram backwards by hand until the indicator plunger bears against the parallel (G) and the indicator needle (I) registers one-quarter of a turn.
   • Set the indicator bezel (J) to zero.

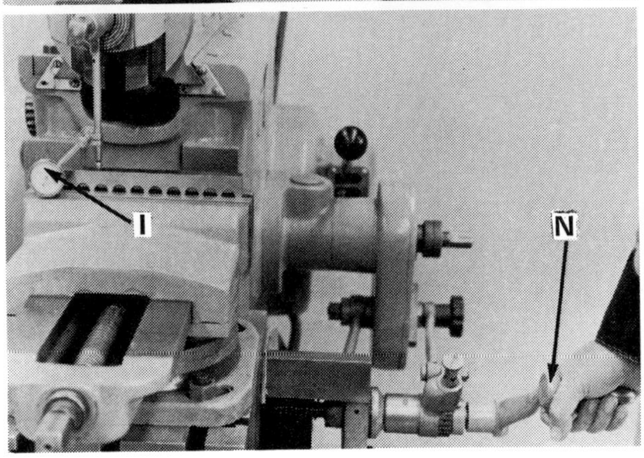

8. • With the crossfeed traverse crank (N) move the table until the indicator bears against the other end of the parallel.
   • Note the difference between both indicator readings.
   • If there is any movement of the needle (I), loosen the swivel base lock nuts slightly.
   • For a high reading at the left-hand end, tap the vise *clockwise* one-half the difference in the two indicator readings.
   • For a high reading at the right-hand end, tap the vise *counterclockwise*.
   • Tighten the swivel base lock nuts and recheck the alignment.

In order to machine a vertical surface, the toolhead must be aligned at 90° to the surface of the table or vise. Two common methods of aligning the toolhead are with a square, and by using an angle plate and a dial indicator.

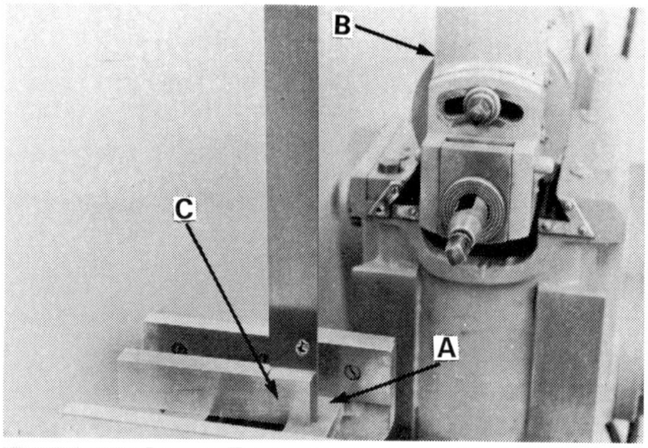

1. **ALIGNING THE HEAD WITH A SQUARE**
   - **Swivel the vise so that the jaws are 90° to the ram travel.**
   - **Clean and remove all burrs from:**
     (1) the base of the vise (A).
     (2) the side of the toolhead (B).
   - **Place a large square (C) on the base of the vise.**

2. - **Move the ram forward until the side of the toolhead (B) is in line with the square (C).**
   - **Hold the beam of the square firmly against the vise base.**
   - **Slide the square forward until the blade of the square (C) lightly contacts the toolhead (B).**
   - **Visually check the alignment of the toolhead and if it is out of alignment, loosen the toolhead lock nut about ¼ of a turn.**

3. - **Place two paper feelers (D) between the blade of the square (C) and the toolhead (B).**
   - **With hand pressure on the beam of the square (C), slide the square until the blade contacts the side of the toolhead (B).**
   - **With a soft-faced hammer, lightly tap the toolhead until both paper feelers are tight.**

4. - **Tighten the toolhead lock (E) firmly.**
   - **Recheck the alignment by using paper feelers between the blade of the square and the toolhead.**
   - **If necessary, readjust the toolhead.**

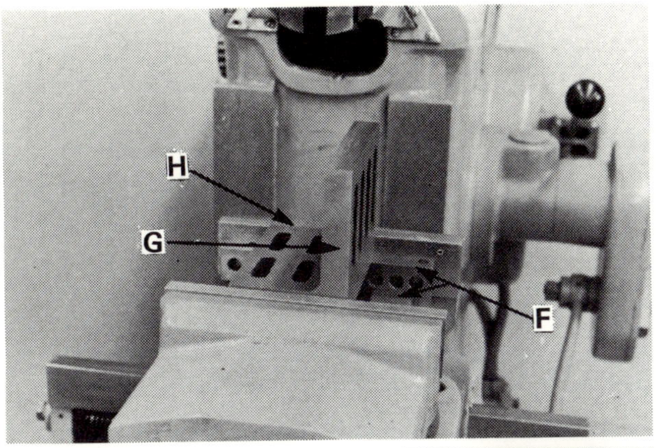

5. DIAL INDICATOR METHOD
   - Clean and remove all burrs from the base of the vise.
   - Align the vise at 90° to the ram.
   - Place a set of parallels (F) in the vise.
   - Set an angle plate (G) on the parallels.
   - Be sure that the edge of the angle plate is against the solid jaw of the vise (H).

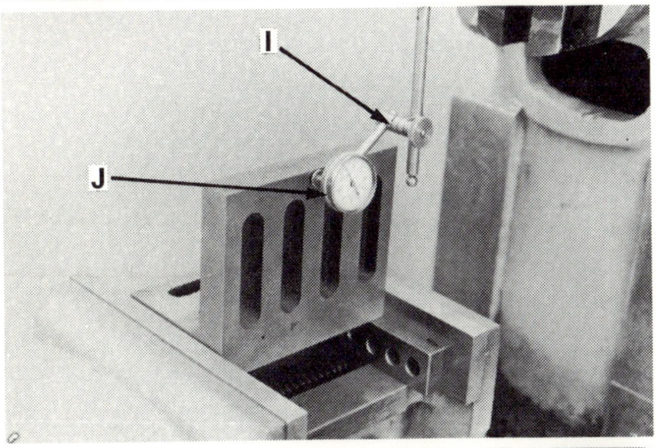

6. - Mount a dial indicator assembly (I) in the toolpost.
   - Check that all parts of the dial indicator assembly will clear the face of the angle plate.
   - Turn the table crossfeed traverse screw until the indicator needle registers about one-quarter of a turn.
   - Turn the bezel (J) by hand until the needle points to zero.

7. - Turn the downfeed handle (K) to move the indicator to the bottom of the angle plate.
   - Note the reading of the indicator. The difference between indicator reading at the top and bottom of the angle plate indicates how much the toolhead is out of line.

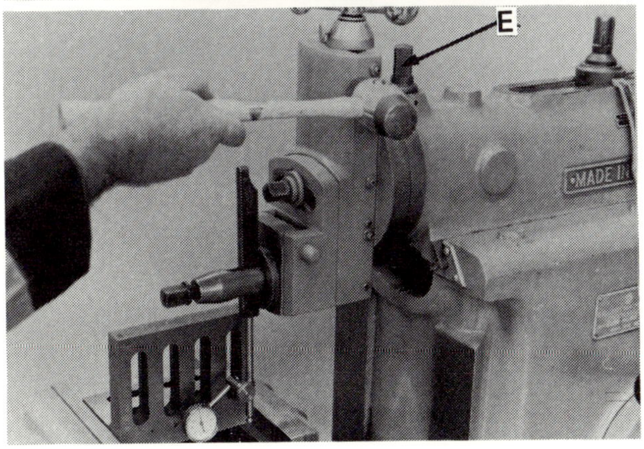

8. - Loosen the toolhead lock (E) about ¼ of a turn.
   - Always adjust the toolhead so the indicator moves away from and *not towards* the angle plate.
   - Tap the toolhead over approximately half the difference between the top and bottom indicator reading.
   - Recheck the surface and adjust the head until the indicator registers zero at the top and bottom.
   - Tighten the toolhead lock (E).

SHAPING A VERTICAL SURFACE

A vertical surface can be machined on flat work by feeding the toolhead with the downfeed handle. In order that the machined edge will be square, the toolhead must be set at 90° to the top of the shaper table.

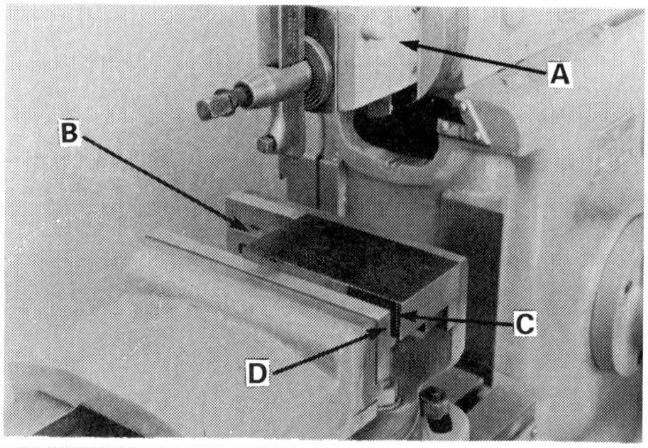

1. • Align the toolhead (A) with a square.
   • Place a set of parallels (B) in the vise.
   • Set the work on parallels so that the layout line (C) extends about ¼" (6.35 mm) past the end of the vise jaw (D) and the parallels (B).
   • Tighten the vise securely and tap the work down on the parallels.
   • Adjust the length and position of the shaper stroke.
   • Set the shaper for the proper speed. (See step 3 Operation 3.)

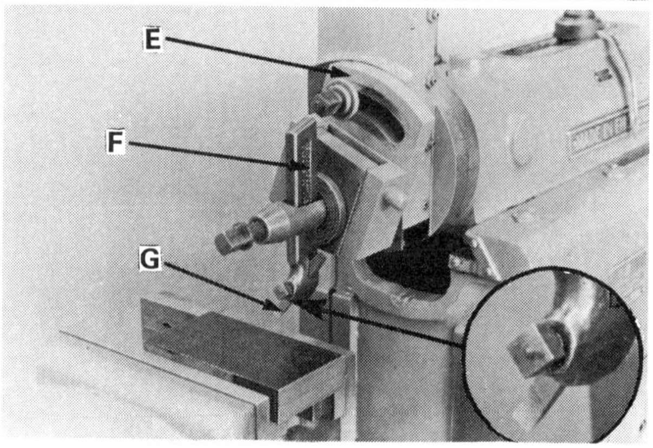

2. • Swivel the top of the apron (E) to the right. This will allow the toolbit to swing away from the work on the return stroke.
   • Fasten a left-hand toolholder (F) in the toolpost.
   • Mount a facing tool (G) in the toolholder. (Be sure that the tool clears the left side of the toolholder.)
   • Raise the toolhead so that the toolbit clears the top of the work by about ½" (12.7 mm).

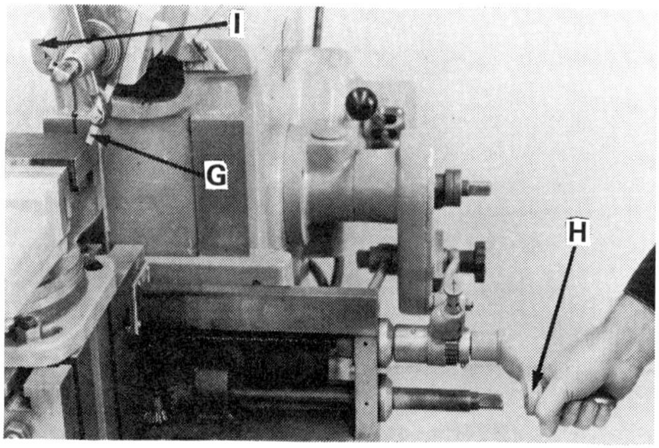

3. • With the crossfeed traverse crank (H), bring the work close to the toolbit (G).
   • Check that the shaper toolhead will not hit the column (I) on the return stroke of the ram.
   • Set the depth of the cut using the crossfeed traverse crank (H).

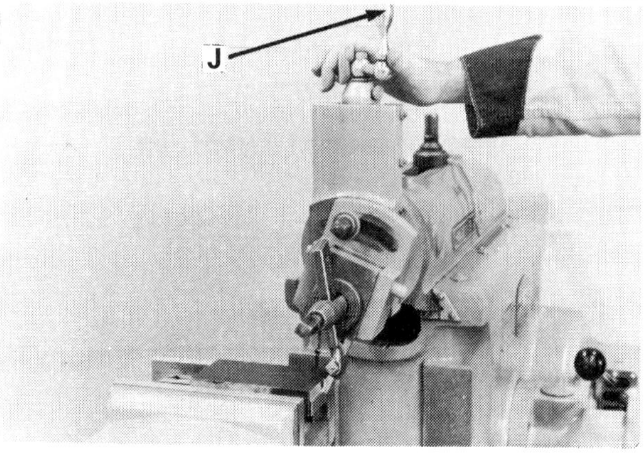

4. • Start the shaper.
   • Feed the toolbit down by turning the downfeed handle (J) clockwise about .010 (.25 mm) on each return stroke of the ram until the surface is cut.
   • Bring the toolbit to the top of the work with the downfeed handle.
   • Set the depth of each cut by using the graduated collar on the crossfeed traverse.

MACHINING AN ANGULAR SURFACE

Angular surfaces can be cut in a shaper by three common methods: (a) setting the workpiece on the desired angle (b) swivelling the vise to the required angle (c) setting the toolhead to an angle.

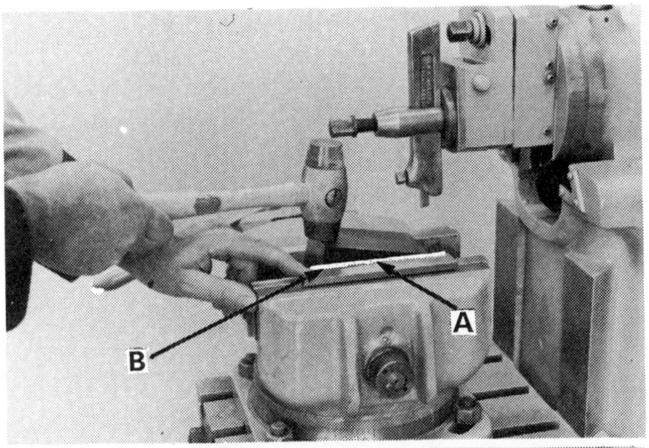

1. SETTING THE WORK ON AN ANGLE
   - Align the vise (Operation #5).
   - Set up the work so that the layout line (A) is parallel with the top of the vise by:
     - (a) laying a parallel (B) on top of the vise.
     - (b) checking the line with a surface gage.
     - (c) measuring with a rule at both ends of the line.
   - Tap the work down until the layout line is in line with the parallel or the top of the vise.
   - Tighten the vise securely.

2. - Set the shaper speed and feed.
   - Set the length and position of the shaper stroke.
   - Bring the toolbit close to the top edge of the work surface.
   - Set the depth of cut using the downfeed handle.
   - Engage the automatic crossfeed (C).
   - Machine to the layout line.

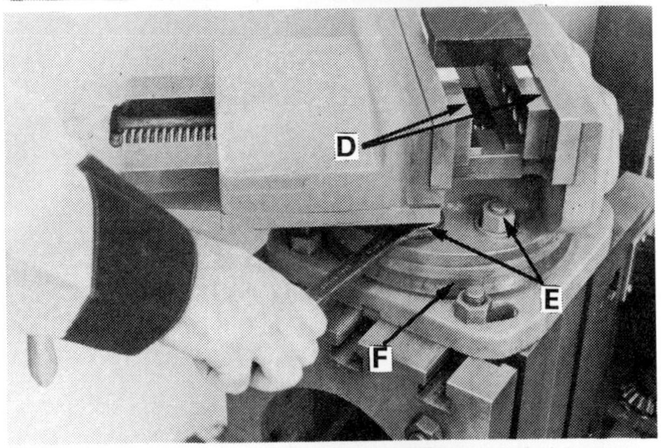

3. SWIVELLING THE VISE
   - Set the work on parallels (D) and in the center of the vise.
   - Loosen the swivel base lock nuts (E).
   - Swivel the vise to the angle required as indicated by the graduations (F) on the base.
   - Tighten the swivel base lock nuts (E).

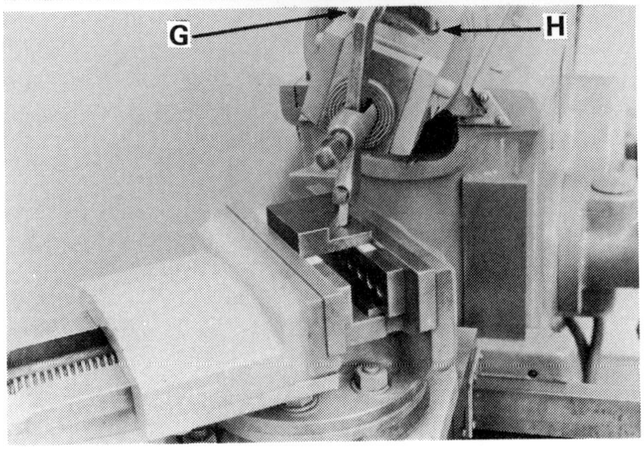

4. - Loosen the apron lock (G).
   - Swing the top of the apron to the right (H) and then tighten the apron lock.
   - Set up a toolholder and cutting tool as shown.
   - Set the depth of cut and engage the table feed.
   - Take successive cuts until the surface is cut to the correct depth and to the layout line.

MACHINING AN ANGULAR SURFACE

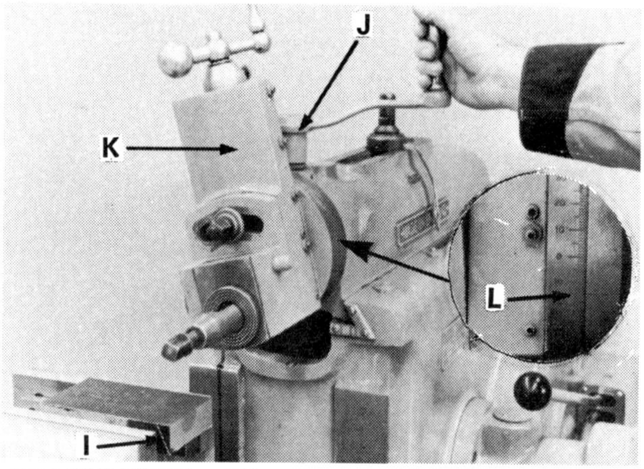

5. SETTING THE TOOLHEAD TO AN ANGLE
    - Mount the work on parallels in the vise.
    - Slide the work until the layout line (I) is about ¼" (6.35 mm) past the edge of the vise.
    - Tighten the vise and seat the work on the parallels.
    - Loosen the toolhead lock (J) slightly.
    - Move the toolhead (K) in the same direction as the layout line (I).
    - Adjust the toolhead to the required angle (L).
    - Tighten the toolhead lock (J).

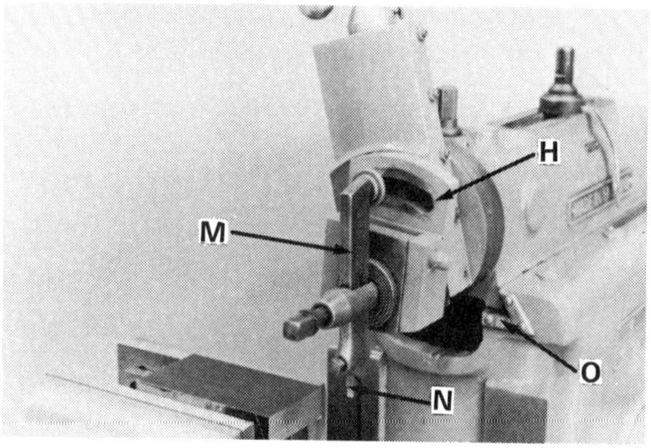

6. - Swivel the top of the apron (H) to the right.
    - Grip a toolholder (M) short in the toolpost.
    - Mount a facing tool (N) in the toolholder.
    - Swivel the toolholder so that its left side will not rub on the work when cutting the angle.
    - Raise the toolhead until the cutting tool clears the top of the work.
    - Check that the toolhead will not hit the column (O) on the return end of the stroke.

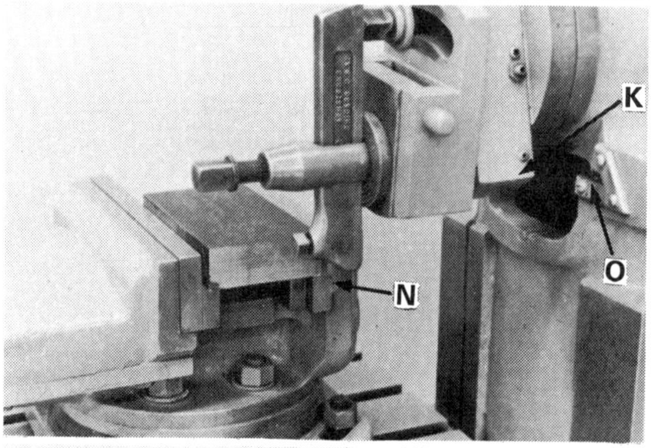

7. - Set the length and position of the shaper stroke.
    - Bring the cutting tool close to the work.
    - Lower the toolhead until the cutting tool (N) just clears the bottom edge of the work.
    - Check that there is clearance between the toolhead (K) and column (O) on the return end of the ram stroke.

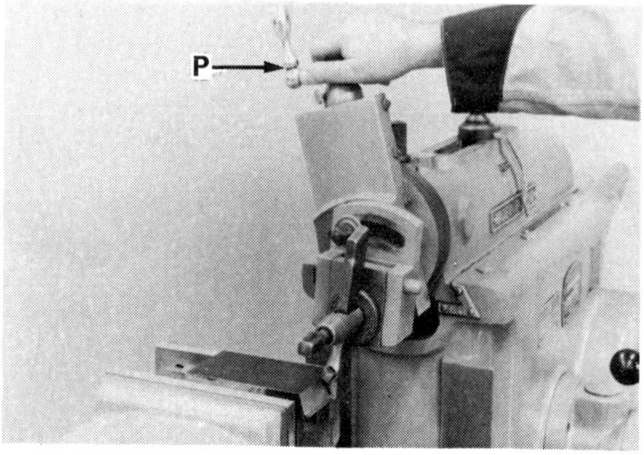

8. - Start the shaper.
    - Feed the toolbit down by turning the downfeed handle (P) clockwise about .010 (.25 mm) on each return stroke of the ram until the surface is cut.
    - Check the angular surface with a protractor.
    - Set the width of each cut by using the graduated collar on the crossfeed traverse screw.

CHANGING A MILLING MACHINE CUTTER

It is often necessary to mount different milling cutters on the arbor in order to perform various milling operations. It is important that the proper procedure be followed to avoid personal injury and damage to the machine and accessories.

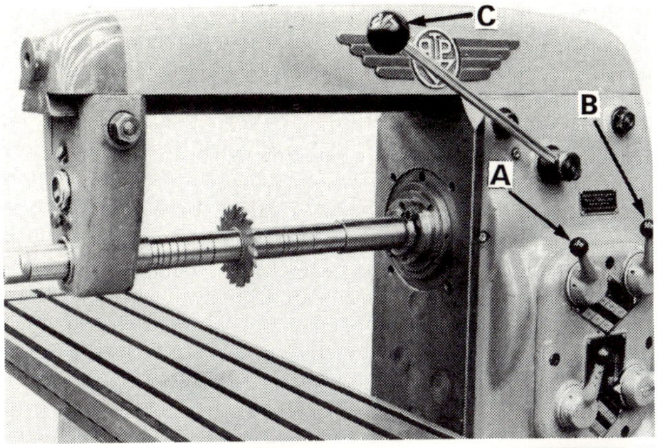

1. • **Shut off the motor of the machine.**
   • **Check the speed chart to find the lowest spindle speed.**
   • **Adjust the levers (A) and (B) to set the machine to the lowest speed.**
   • **Engage the spindle clutch lever (C).**

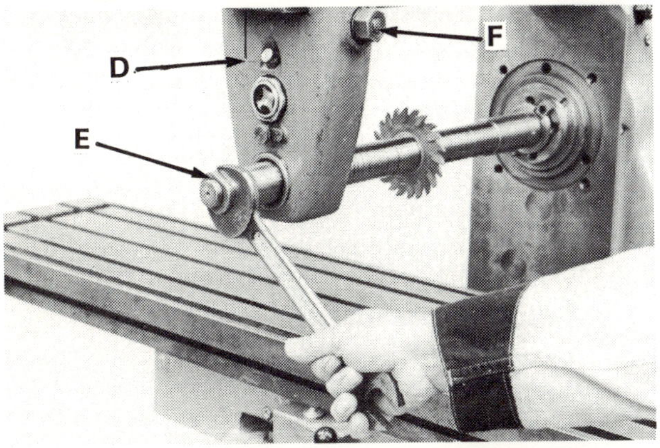

2. • **Be sure the arbor support (D) is in position and supporting the arbor on the bearing bushing.**
   • **Loosen the arbor nut (E).**
   NOTE: If the spindle has a right-hand thread, turn the nut counterclockwise to loosen; left-hand threads loosen clockwise.
   • **If the nut is tight, give the wrench a sharp pull or strike it sharply in the proper direction with the hand.**
   • **Loosen the arbor support nut (F) ½ turn.**

3. • **Place a piece of masonite (G) on the milling machine table. (This will protect the table surface from becoming scratched or damaged.)**
   • **Using both hands, pull the arbor support (D) off the overarm (H).**
   • **Place the arbor support on the masonite.**

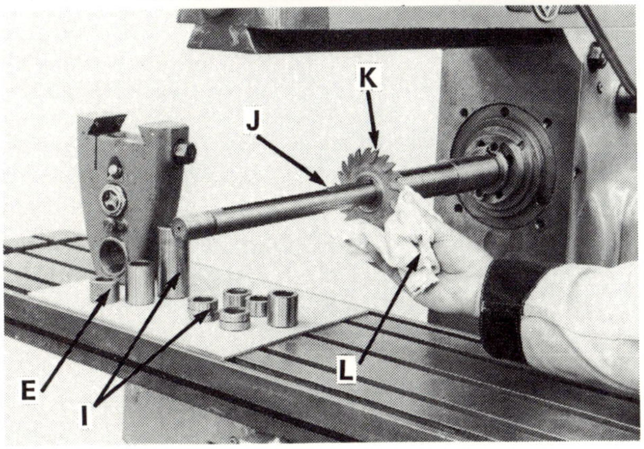

4. • **Remove the arbor nut (E) and place it on the masonite.**
   • **Remove the arbor collars (I) and key (J) from the arbor and place them on the masonite.**
   • **Hold the milling cutter (K) with a cloth (L) and remove it from the arbor.**
   • **Wipe off all the chips and store the cutter in a cupboard.**

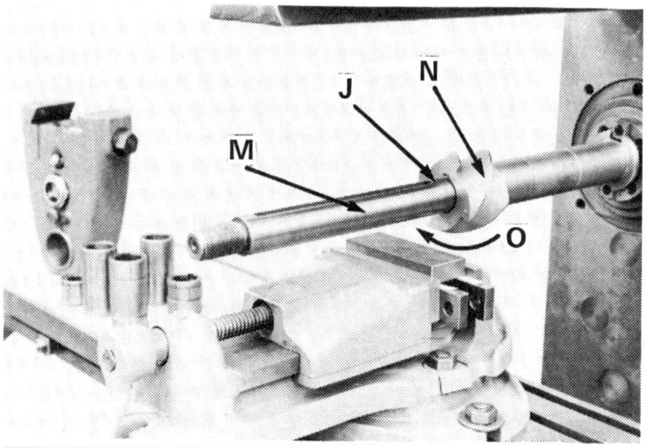

5. • Clean the arbor (M) and the cutter faces with a cloth.
   • Mount the proper cutter (N) on the arbor using a cloth to protect the hands. NOTE: Be sure that the cutter teeth are pointing in the direction the spindle rotates (O).
   • Align the keyways of the arbor and the cutter.
   • Insert the key (J) in the keyway.
   NOTE: Key all cutters except slitting saws.

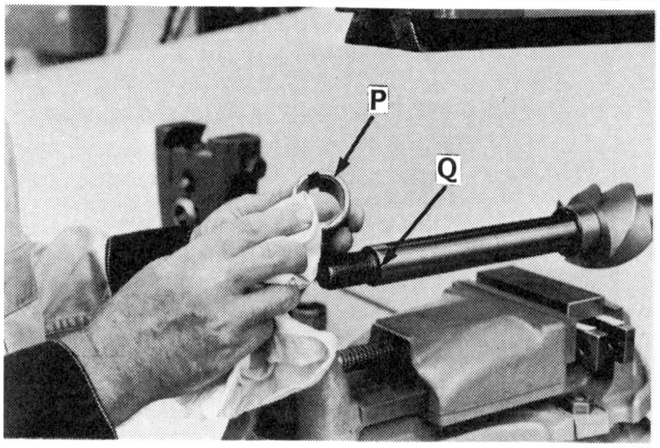

6. • Clean both ends of all spacer collars (P) and mount them on the arbor. It is important that all bushings be free of burrs and chips.
   • Mount the bearing bushing (the one with the largest diameter) on the arbor as close to the cutter as possible.
   • Mount the required number of spacer collars to just project beyond the arbor shoulder (Q).
   • Replace the arbor nut and tighten it *by hand.*

7. • Oil the bearing bushing (R).
   • Slide the arbor support (D) on the overarm (H) and position it over the bearing bushing.
   • Tighten the arbor support nut (F) to lock the support in position.
   • Check the oil level in the sight-glass (S) and refill if necessary.

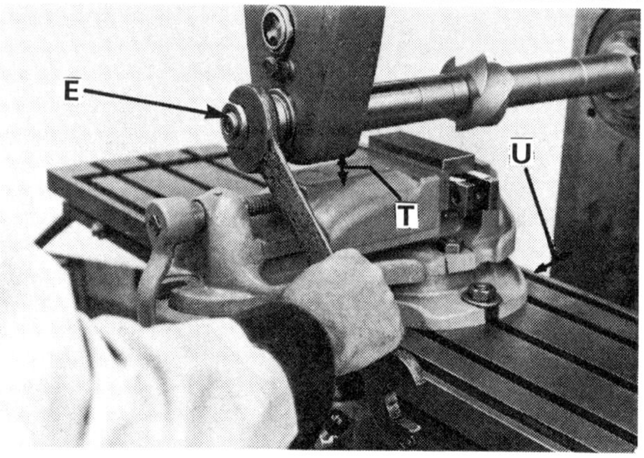

8. • Tighten the arbor nut (E) securely with a wrench. (A left-hand thread tightens counter-clockwise; a right-hand thread tightens clockwise.)
   • Raise the table to bring the work to within about 1/8" (3.17 mm) of the cutter.
   • Check the distances (T) and (U) to ensure sufficient clearance at each point when the work is being machined.

CHANGING AN ARBOR

Milling machine cutters are supplied with various size holes. When a different cutter is required for a job, it may be necessary to change the arbor to fit the hole size of the cutter.

1. • Stop the motor.
   • Remove the cutter as in frames 1 to 4 of Operation #1, "Changing a Milling Cutter".
   • Clean the arbor, key, and all collars.
   • Replace the key, collars and arbor nut on the arbor.
   • Tighten the arbor nut (A) hand tight.

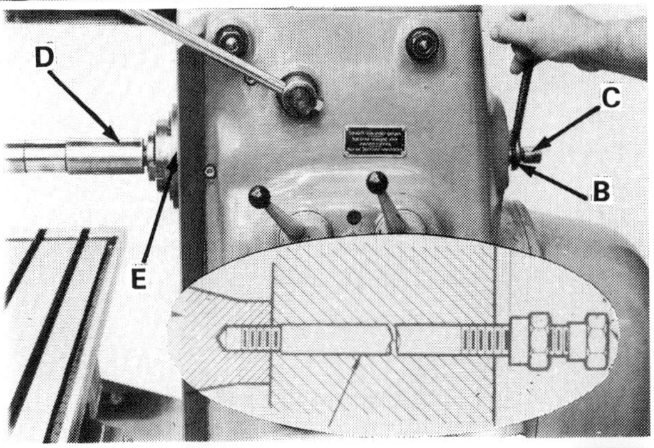

2. • With the proper wrench, loosen the draw-in bar lock nut (B) a few turns.
   • The draw-in bar (C) threads into the end of the arbor (D) and holds it in the tapered hole of the spindle (E).
   • Unscrew the draw-in bar two or three turns.

3. • Hold the arbor (D) with one hand.
   • With a soft-faced hammer (F) sharply strike the end of the draw-in bar (C). This will release the fit between the taper on the arbor (D) and the taper in the spindle (E).

4. • Hold the arbor (D) with one hand.
   • Unscrew the draw-in bar (C) from the arbor with the other hand.
   • Carefully remove the arbor from the spindle taper and place it on a piece of masonite on the table.
   • Leave the draw-in bar (C) in the machine spindle for use with the replacement arbor.

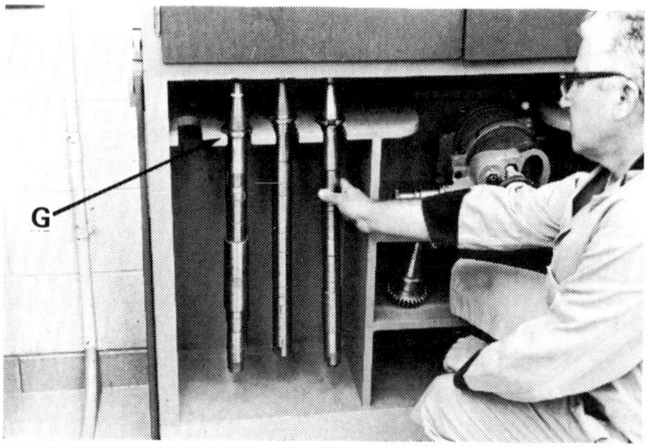

5. • Clean and apply a light film of oil to the arbor.
   • Store it in a proper wooden rack (G) to avoid damage to the taper and bushings. (Hanging arbors upright will prevent them from warping or bending.)

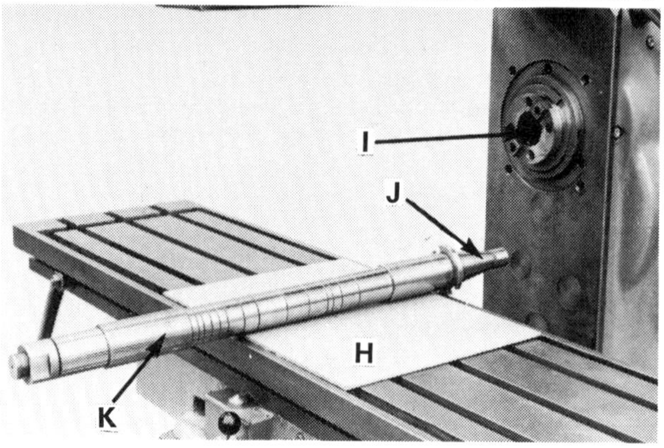

6. • Select the required arbor to fit the cutter to be used, and place it on a piece of masonite (H) on the milling machine table.
   • Clean the tapered hole (I) in the end of the spindle.
   • Clean the tapered end of the arbor (J) and remove any burrs with a honing stone.
   • Inspect the bearing bushing (K) and remove all burrs with a honing stone.

7. • Carefully insert the tapered end of the arbor into the milling machine spindle.
   • Align the slots (L) on the arbor collar with the driving lugs (M) on the milling machine spindle.
   • Push the arbor into the spindle as far as possible.
   • Thread the draw-in bar (C) into the arbor until the shoulder of the jam nut (B) is about ⅛″ away from the spindle shoulder.
   • Tighten the jam nut (B) by hand up to the spindle shoulder.

8. • With the proper wrench, tighten the jam nut (B) against the spindle shoulder.
   • Set the speed change levers (N) to about 100 rpm.
   • Start the motor and engage the clutch lever (O).
   • Check that the arbor is running true.
   • If any runout is noticeable, shut off the machine and remove the arbor.
   • Recheck the taper and tapered hole for burrs or metal particles and replace the arbor.

ALIGNING THE VISE

In order that work held in a vise be machined accurately, the vise must be properly seated on the milling machine table and aligned with the table travel.

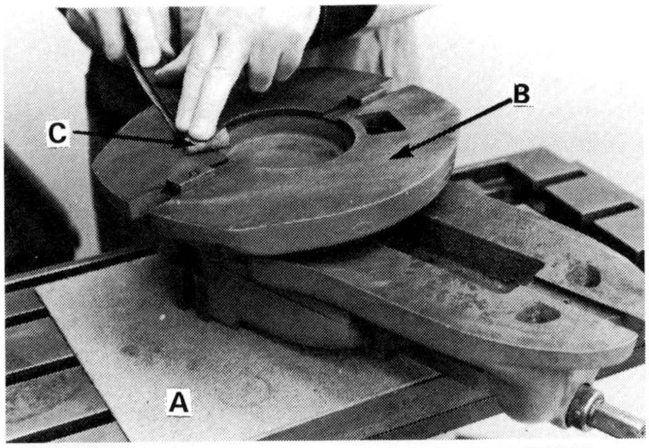

1. • Place the vise on a piece of masonite (A) and examine the base for burrs and dirt.
   • Clean the bottom of the base (B).
   • Remove all burrs with a piece of abrasive cloth (C) folded over the end of a flat file.

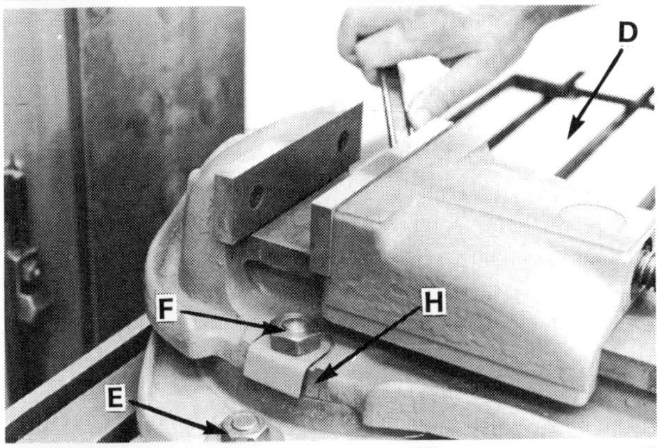

2. • Clean the milling machine table (D) and remove any burrs with a honing stone.
   • Mount the vise on the table with the base keys in the center slot of the table.
   • Fasten the vise to the table by tightening the nuts (E).
   • Loosen the swivel base lock nuts (F) and swivel the vise until the indicator mark (H) is aligned with the 90° mark.
   • Lightly tighten the lock nuts (F).

3. • Loosen the arbor nut and place the dial indicator assembly bar (G) between two collars (H).
   • Tighten the arbor nut only enough to hold the indicator assembly in place.
   • Loosen the lock nut (I) and adjust the indicator so that the plunger (J) points toward the solid jaw of the vise (K).
   • Tighten the lock nut (I) securely.

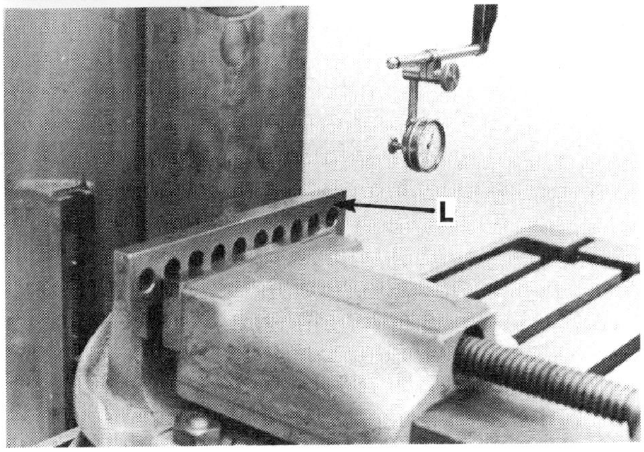

4. • Clean the vise jaws and remove any burrs with a honing stone.
   • Tighten a clean parallel (L) in the vise with about 1" (25.4 mm) projecting above the jaws.

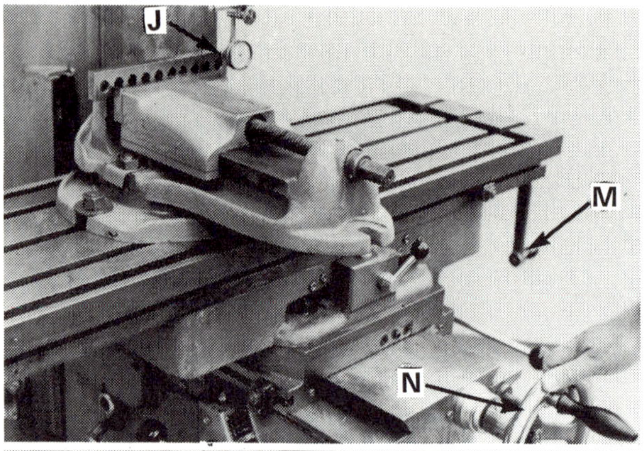

5. • Raise the table until the indicator plunger (J) is about ⅜″ (9.52 mm) below the top of the parallel.
   • Adjust the table with the longitudinal feed handle (M) and the crossfeed handle (N) until the indicator plunger is close to the right-hand end of the parallel.

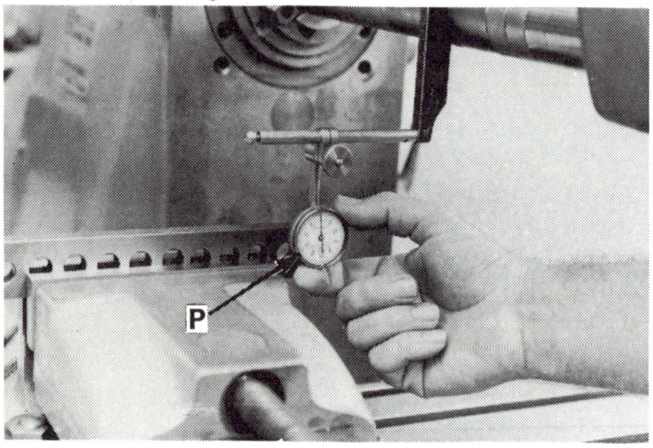

6. • Turn the crossfeed handle until the plunger bears against the parallel and the indicator needle registers about one-quarter of a revolution.
   • Tighten the saddle clamp to lock the table in position.
   • Turn the bezel (P) until the indicator registers zero.

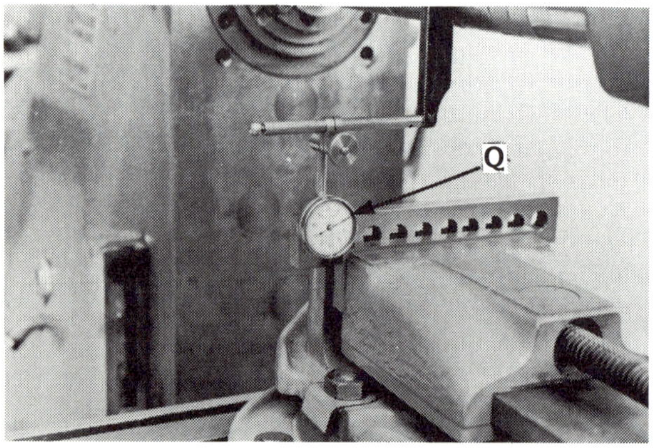

7. • Turn the longitudinal feed handle and move the table until the indicator is at the left-hand end of the parallel.
   • Note the reading on the dial indicator (Q).
   • Any difference in the indicator reading between the right-hand and left-hand end of the parallel would indicate that the vise is out of alignment.

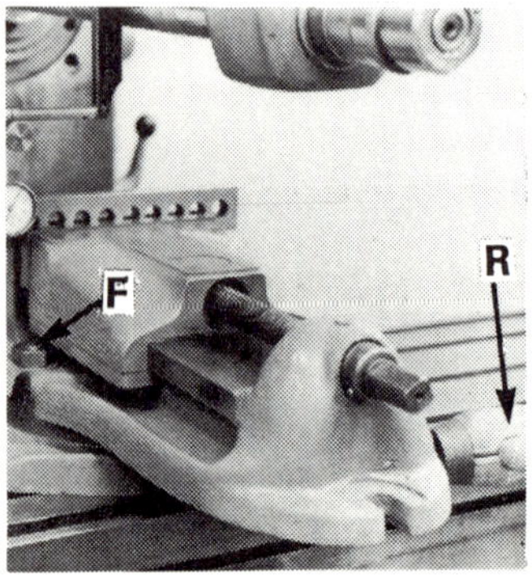

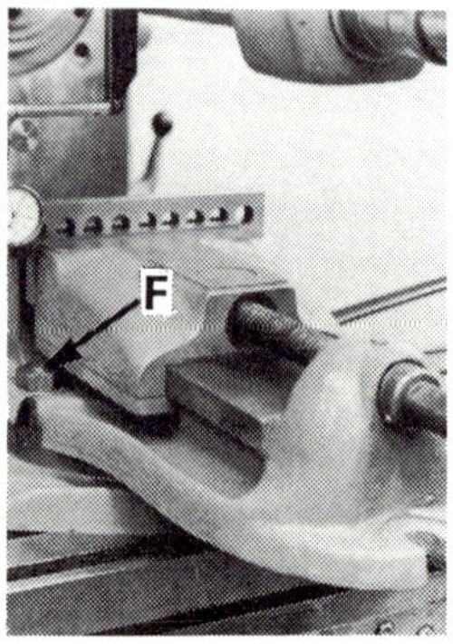

8. • Loosen the locking nuts (F) slightly.
   • Tap the vise with a soft-faced hammer (R) until the needle registers *half the difference* between the two indicator readings.
   • When the indicator registers high at the left-hand end, tap the vise clockwise; for a high reading at the right-hand end, tap the vise counterclockwise.

     Always tap the vise so that the parallel moves *away* from the indicator plunger.
   • Tighten the lock nuts (F).
   • Recheck the indicator reading at both ends of the parallel.

MILLING A FLAT SURFACE

Flat surfaces are generally machined on workpieces with a helical plain milling cutter. The work may be held in a vise or clamped directly to the table.

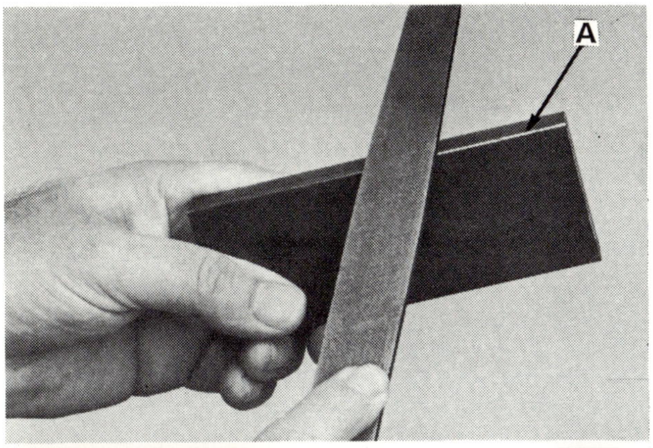

1. • **Remove the burrs from all edges of the workpiece (A) with a mill file.**
   • **If all burrs are not removed from a workpiece, they:**
   (a) **prevent the work from seating properly.**
   (b) **cause the work to tip in the vise.**
   (c) **cause the surface to be machined out-of-square.**

2. • **Clean the workpiece and place it on a piece of masonite (B) on the table.**
   • **Move the table until the vise is well clear of the cutter.**
   • **Thoroughly clean and wipe the vise jaws and throat.**
   • **Remove any burrs from the vise jaws with a honing stone.**

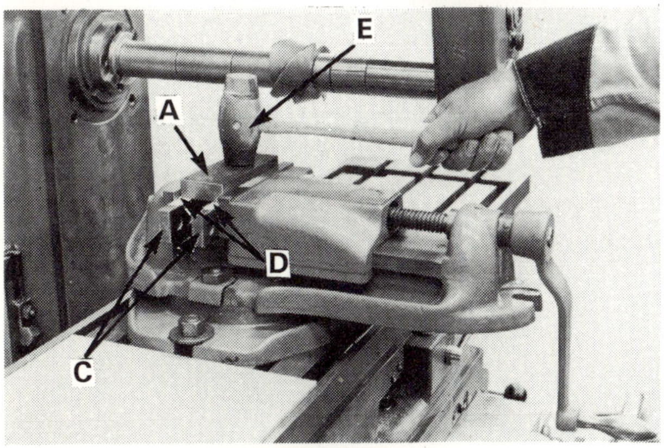

3. • **Select a pair of parallels (C) which will raise the work (A) about ¼″ (6.35 mm) above the vise jaws.**
   • **Set the work on the parallels and in the center of the vise.**
   • **Place a short paper feeler (D) between each corner of the work and the parallels.**
   • **Tighten the vise securely.**
   • **Tap the work down with a soft-faced hammer (E) until all paper feelers are tight.**

4. • **Calculate the proper speed for the cutter.**

$$\text{rpm} = \frac{4 \times \text{CS}}{\text{Cutter Dia.}}$$

   CS is cutting speed of the material.
   • **Locate the closest rpm on the speed chart (F).**
   • **Set the speed-change levers (G) and (H) to the proper positions.**
   • **If it is difficult to engage one of the levers, turn the arbor (I) by hand until the lever engages.**

5. • Calculate the proper feed for the work material and the finish required.
   Feed = rpm × number of teeth in cutter × feed per tooth
   • Set the levers (J) and (K) to the positions indicated on the feed chart (L).

A B

6. • Fig. 6A illustrates climb or down milling. The cutter revolves down onto the work in the direction (M) and the work is fed into the cutter in the direction (N).
   NOTE: When climb milling, always use a table backlash eliminator.
   • Fig. 6B shows conventional or up milling. The cutter revolves in the same direction (M) as for climb milling. The work is fed into the cutter from the opposite end in the direction (O).

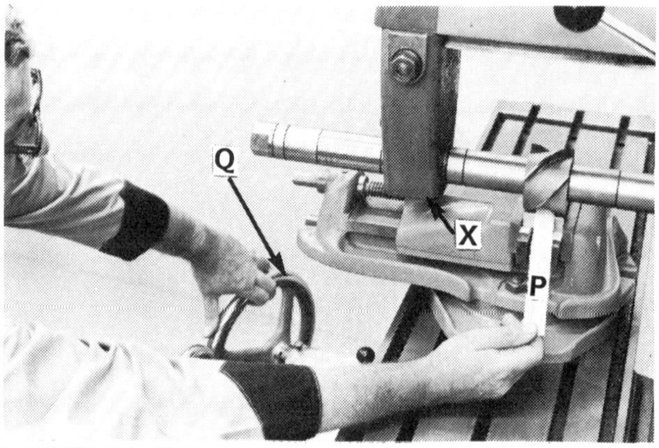

7. • Start the cutter and bring the end of the work under the cutter.
   • Place a *long strip* of paper (P) on top of the workpiece and grip it *lightly* by hand.
   • Turn the vertical feed handwheel (Q) to raise the table until the cutter just cuts the paper.
   • Check that there is clearance (X) between the arbor support and the vise.

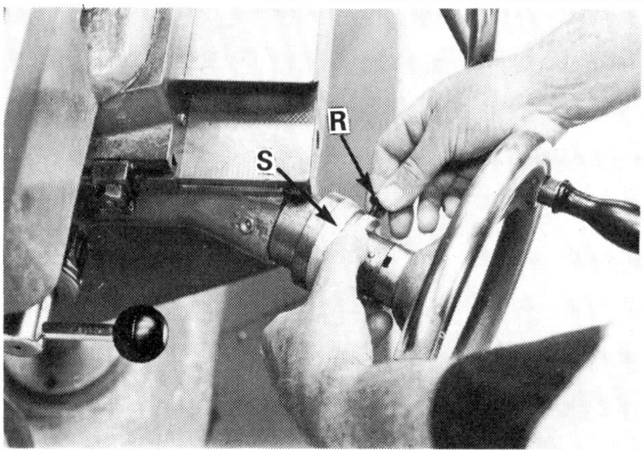

8. • Stop the machine.
   • Loosen the set screw (R).
   • Set the vertical feed graduated collar (S) to zero.
   • Tighten the set screw (R).

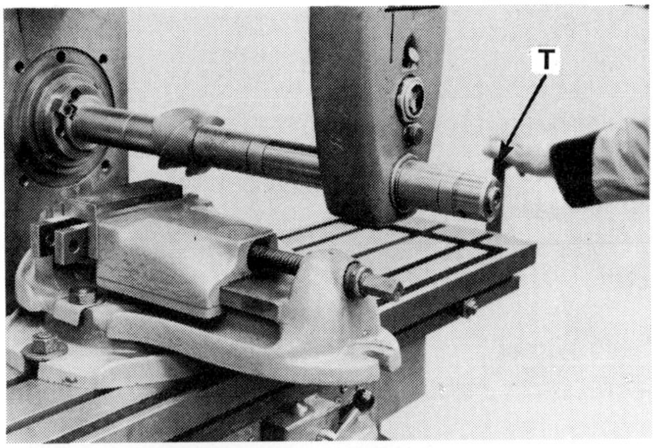

9. • Turn the table feed handle (T) until the end of the work is about ½" (12.7 mm) from the cutter.
   • Adjust the saddle so that the work is in the center of the cutter width.
   • Clamp the saddle to prevent any side movement.

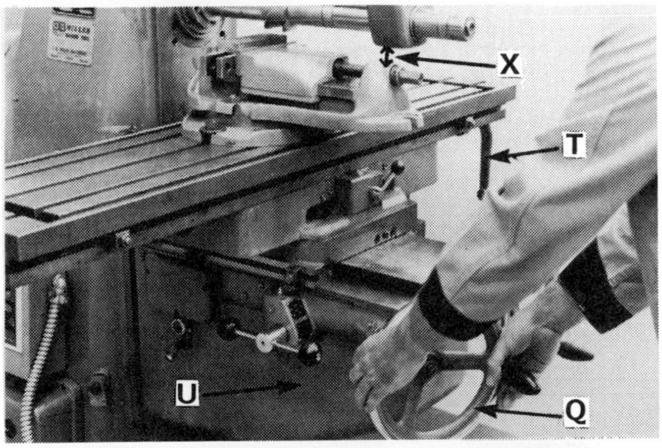

10. • Turn the vertical feed handwheel (Q) clockwise to set the desired depth of cut.
   • Lock the knee (U) at this height.
   • Start the cutter revolving.
   • Turn the table feed handle (T) until the end of the work just touches the revolving cutter.
   • Check the clearance (X) between the arbor support and the vise.

11. • Engage the automatic feed lever (V).
   • Apply cutting fluid to the cutter and work.
   • Take a trial cut about ¼" (6.35 mm) long.
   • Disengage the automatic feed (V).
   • Stop the cutter from revolving.

12. • Move the work clear of the cutter with the table feed handle.
   • Measure the thickness of the work with a micrometer (W).
   • If the work is too thick, unlock the knee clamp and raise the table the desired amount.
   • Lock the knee clamp and machine the work to size.

MILLING A VERTICAL SURFACE

Vertical surfaces can be machined when held in a vise by using a side milling cutter. The thickness of the work that can be machined depends upon the diameter of the milling cutter.

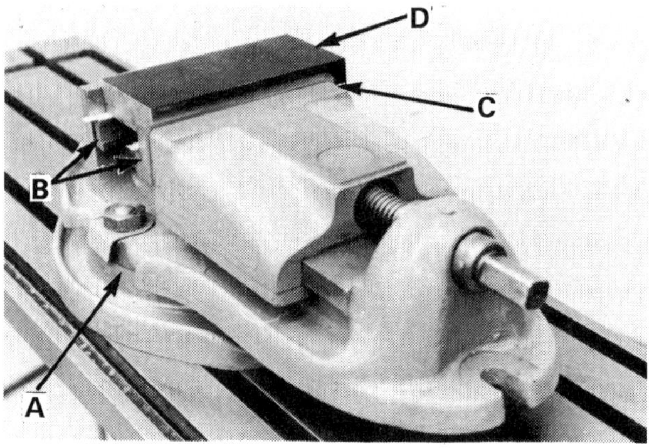

1. • Swivel the vise so that the jaws are at 90° to the table slot. This is indicated on the graduated scale (A) of the base plate.
   • Clean and align the vise with a square or dial indicator. (See Operation #3.)
   • Place parallels (B) in the vise. Do not extend them beyond the end of the vise (C).
   • Set the work on parallels so that the end (D) extends about ½" (12.7 mm) past the vise jaws.
   • Tighten the vise jaws securely and seat the work on the parallels.

2. • Move the table within 2" (50.8 mm) of the column face (E) with the crossfeed handle.
   • Mount a side milling cutter (F) as close as possible to the spindle (G) with its edge in line with the end of the work.
   • Fasten the arbor support (H) close to the milling cutter.
   • Set the machine for the proper speed and feed.

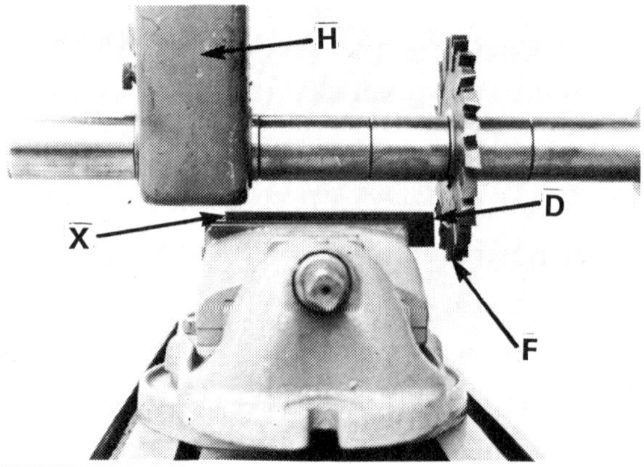

3. • Move the table to bring the end of the work (D) close to the cutter (F).
   • Advance the table to bring the work under the center line of the arbor.
   • Raise the table until the cutter clears the bottom of the work.
   • Check for clearance (X) between the bottom of the arbor support (H) and the work.

4. • Set the proper speed and feed.
   • Start the cutter revolving.
   • Move the table lengthwise until the right-hand edge (I) of the work overlaps the cutter by about ¼" (6.35 mm).
   • Turn the table crossfeed handle until the revolving cutter just touches the layout line (J).
   • Lock the saddle in position.
   • Engage the table automatic feed and machine the edge.

MILLING AN ANGULAR SURFACE

Angular surfaces can be machined on a horizontal milling machine by three methods: (a) Setting the workpiece on an angle, (b) Swivelling the vise, (c) Using an angular milling cutter.

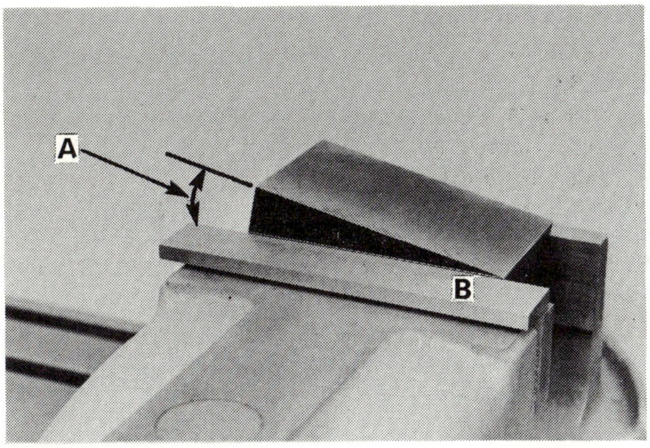

1. **SETTING WORK ON AN ANGLE**
   (A) *Parallel Method*
   - Lay out the required angle (A) on the workpiece.
   - Clean the vise and work and remove all burrs.
   - Mount the work with the layout line (A) about ¼" (6.35 mm) above the vise jaws.
   - Place a parallel (B) on the vise jaw.
   - Tap the work until the layout line (A) is in line with the top edge of the parallel.
   - Tighten the vise securely and recheck the layout line.

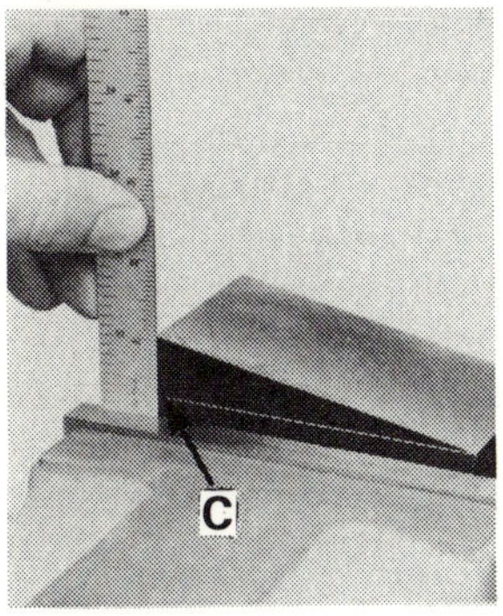

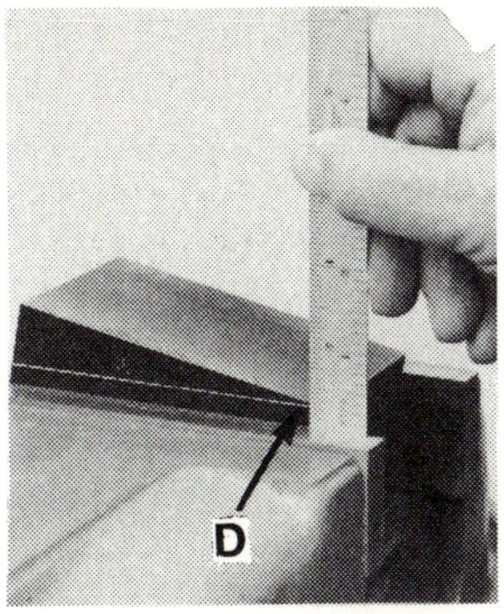

2. (B) *Rule Method*
   - Place the work in the vise with the layout line about ¼" (6.35 mm) above the vise jaws.
   - Lightly tighten the vise enough to hold the work in position.
   - Measure the distance from the top of the vise jaw to the line at each end of the work.
   - Tap the work down until the measurements at (C) and (D) are the same.
   - Tighten the vise securely and recheck the height of the line at each end of the work.

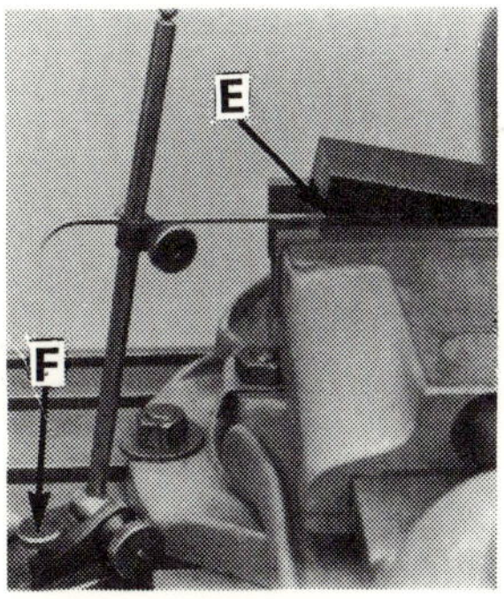

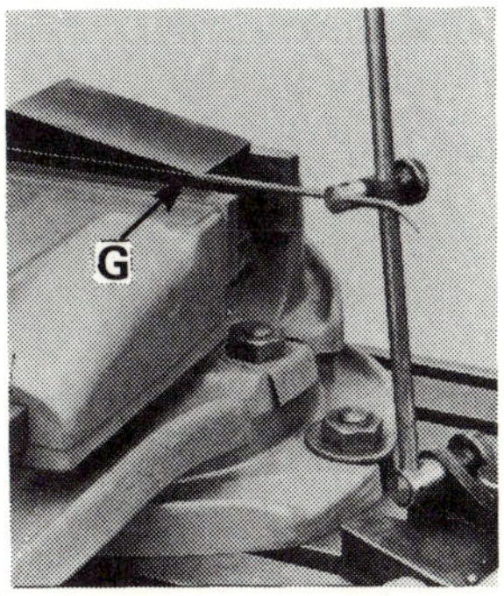

3. (C) *Surface Gage Method*
   - Lightly tighten the vise so that the layout line (E) is approximately parallel to the top of the vise jaw.
   - Set the surface gage scriber point to the line (E) by turning the adjusting screw (F).
   - Move the surface gage to the other end (G) and check the height of the line.
   - Tap the work until the line is the same height as the scriber point at both ends.
   - Tighten the vise securely and recheck the layout line at both ends.

4. **MILLING AN ANGLE**
   - Mount a plain milling cutter on the arbor.
   - Set the proper speed and feed.
   - Start the cutter and raise the table until work touches the cutter.
   - Move the work clear of the cutter and raise the table for the depth of cut desired.
   - Check the clearance (X) between the vise and the arbor support.
   - Rough mill to about 1/32" (.79 mm) above the layout line.
   - Raise the table and take the finish cut to the layout line.

5. SWIVELLING THE VISE
   - Clean and remove all burrs from the work and vise.
   - Place the work on suitable parallels (H) so that the finished depth will be about 1/4" (6.35 mm) above the vise jaws.
   - Fasten the work securely in the center of the vise.
   - Tap the work down on the parallels.
   - Loosen the swivel base lock nuts (I).
   - Swivel the vise to the required angle.
   - Tighten the lock nuts and recheck the setting.

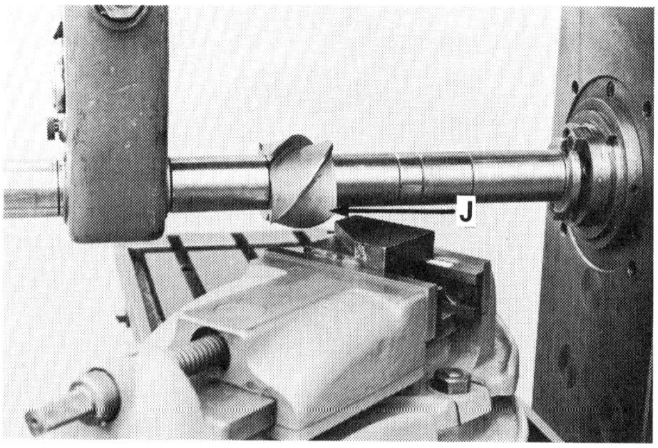

6.
   - Mount the proper cutter on the arbor.
   - Raise and adjust the table until the work is about 1/8" (3.17 mm) below the cutter and the layout line is directly below the edge (J) of the cutter.
   - Move the work clear of the cutter and start the cutter revolving.
   - Raise the table the required amount and lock the knee clamp.
   - Engage the automatic feed lever and take the cut.

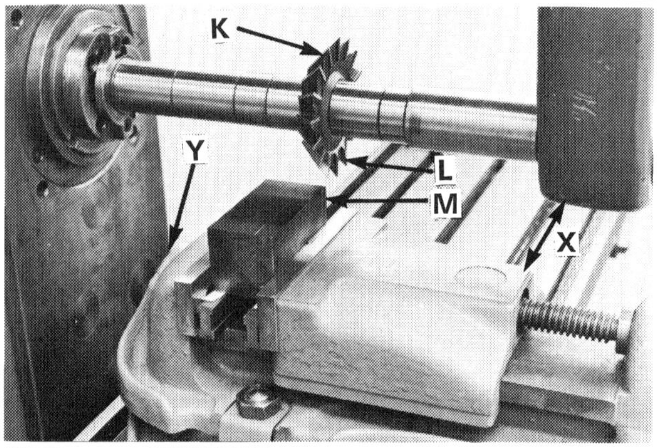

7. USING AN ANGULAR CUTTER
   - Remove all burrs and clean the vise and work.
   - Align the solid jaw of the vise.
   - Seat the work properly in the vise.
   - Mount a single angle cutter (K) on the arbor.
   - Check for clearance at (X) and (Y).
   - Adjust the table so that the face (L) of the cutter overlaps the edge (M) of the work by about 1/8" (3.17 mm).
   - Move the table longitudinally until the work is clear of the cutter.

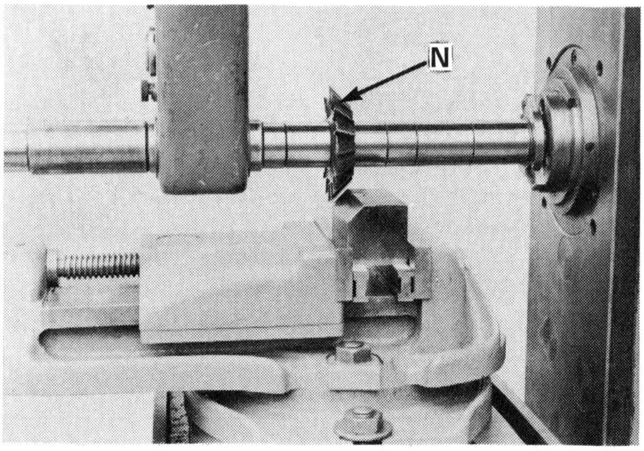

8.
   - Set the speed and feed for the cutter and material.
   - Start the cutter and raise the table until the cutter face (N) cuts to within 1/32" (.79 mm) of the layout line.
   - Engage the automatic feed and take the roughing cut.
   - Return the table, raise the table the required amount and take the finish cut.

MOUNTING A CUTTER

A variety of cutters and end mills can be used in a vertical milling machine for performing various operations. These cutting tools can be held in the machine spindle by spring collets, collets and adaptor, or solid collets.

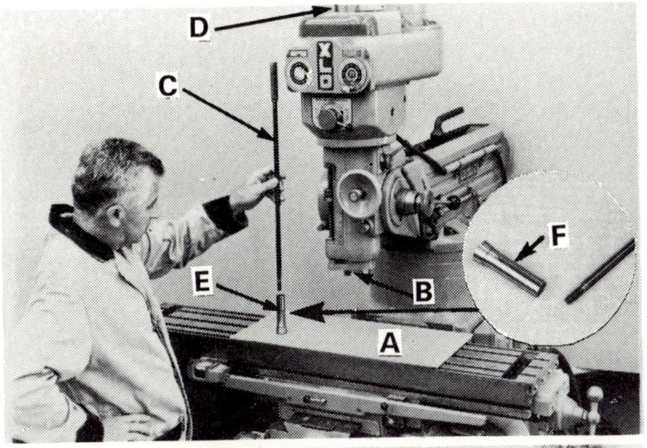

1. **MOUNTING AN END MILL IN A COLLET**
   - Place a piece of masonite (A) on the table.
   - Clean the inside of the spindle taper (B).
   - Slide the draw bar (C) through the top of the spindle (D).
   - Clean the collet taper (E) and keyway (F).
   - Hold the collet up in the spindle (B) and turn it slowly until the keyway (F) fits into the key of the spindle.
   - Hold the collet (E) and thread the draw bar (C) into the collet.

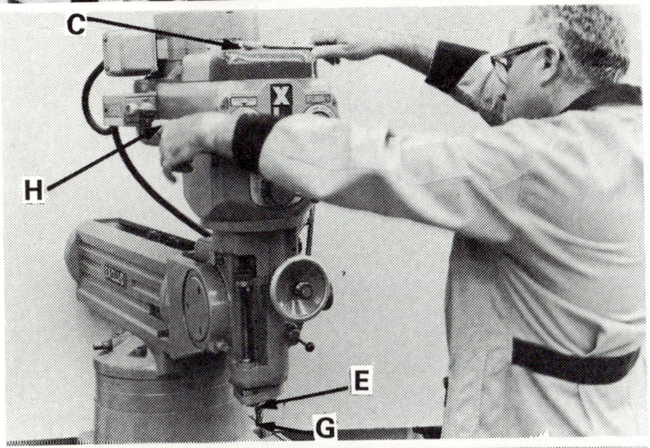

2. - Hold the end mill (G) with a rag and insert it into the collet (E).
   - Tighten the draw bar into the collet by hand.
   - Hold the spindle brake lever (H) with one hand.
   - Tighten the draw bar (C) with a wrench while holding the spindle brake lever (H).

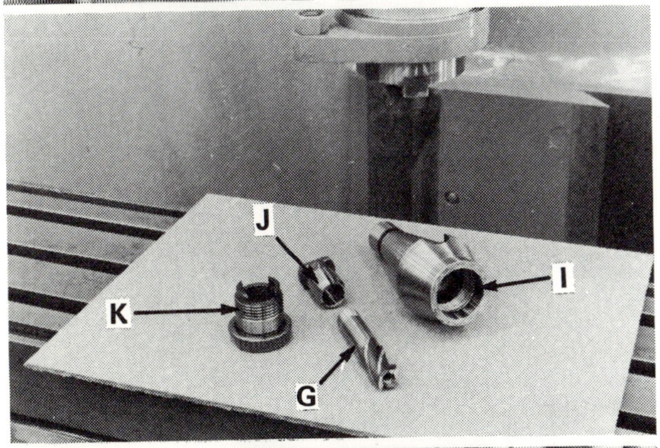

3. **COLLETS AND ADAPTOR**
   - Remove all burrs from the adaptor (I) and the shank of the end mill (G).
   - The adaptor (I) is held in the spindle with a draw bar.
   - The collet (J) holds the end mill (G).
   - The retainer nut (K) tightens the end mill into the collet and the adaptor.

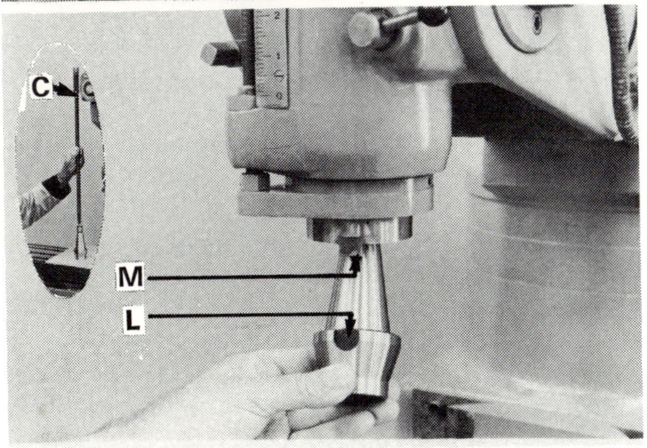

4. **MOUNTING AN END MILL IN AN ADAPTOR**
   - Align the slots (L) of the tapered adaptor with drive keys (M) of the spindle.
   - Hold the adaptor up in the spindle.
   - Thread the draw bar (C) into the adaptor by hand.
   - Hold the spindle brake lever and tighten the draw bar with a wrench.

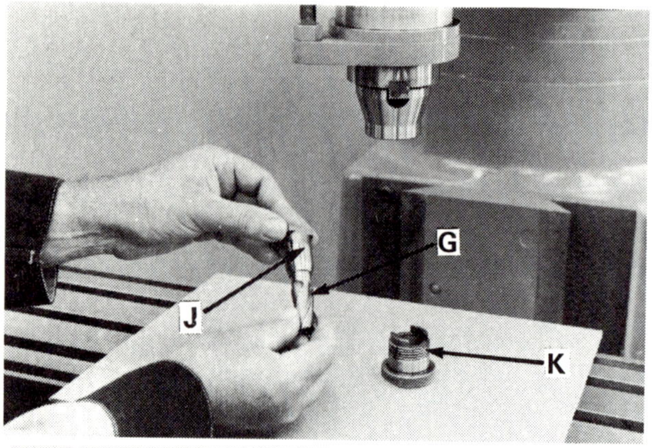

5. • Clean the inside of the collet.
   • Thread the end mill (G) into the collet (J) two or three turns.
   • Clean the outside of the collet and the inside of the retainer nut (K).
   • Slide the retainer nut (K) on the collet.

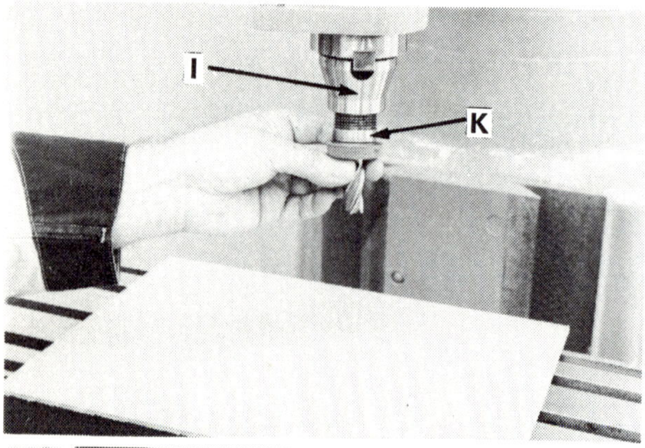

6. • Clean the hole of the adaptor (I).
   • Fit the assembled end mill into the adaptor.
   • Thread the retainer nut (K) all the way into the adaptor (I) by hand.

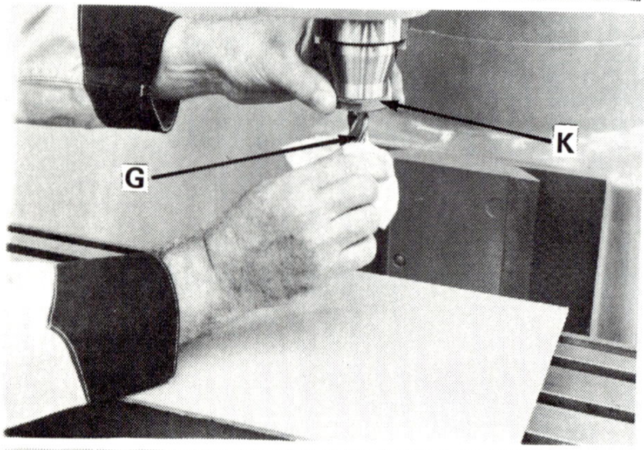

7. • Hold the retainer nut (K) with one hand.
   • Use a rag to thread the end mill (G) into the collet by hand.

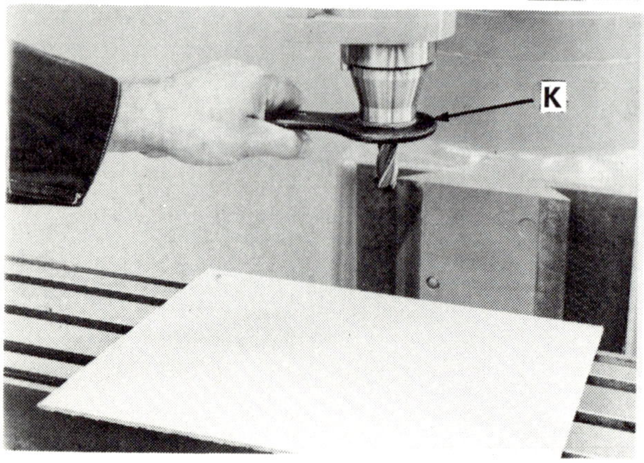

8. • Hold the spindle brake lever with one hand.
   • Tighten the retainer nut (K) securely.
   • Start the machine and check the end mill for trueness.
   • If it is not running true, remove the adaptor and inspect each part for burrs or dirt.

ALIGNING THE HEAD

The alignment of the head is very important when machining holes, surfaces at right angles, or face milling. If the head is not at right angles (90°) to the table, the holes or surfaces machined will not be square or accurate.

1.
- Use the graduations on the head (A) and set to zero.
- Clean and remove all burrs from the surface of the table.
- Clean the machine spindle and mount a drill chuck (B).
- Fasten a ½" (12.7 mm) diameter rod (C) in the chuck.
- Have the rod projecting about 2" (50.8 mm) from the chuck.

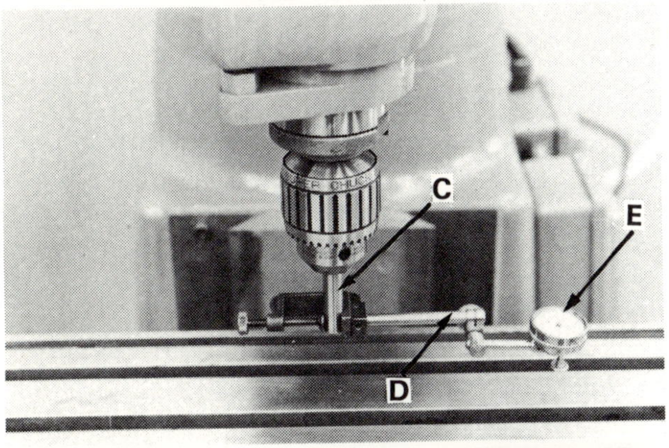

2.
- Fasten a dial indicator assembly (D) on the round rod (C).
- Fasten the indicator (E) near the end of the assembly.
- Position the indicator so that its plunger is at right angles (90°) to the table.

3.
- Turn the drill chuck by hand until the indicator is pointing to the right.
- Move the table with the crossfeed handle until the indicator is in the middle of the table.
- Raise the table until the indicator registers about one-quarter of a turn.

4.
- By hand, rotate the spindle in a small arc to find the highest indicator reading at the right-hand side of the table.
- Keep the indicator plunger clear of the table slot.
- Stop when the indicator is at the highest reading.
- Set the indicator bezel (F) to zero.

5. • Rotate the spindle by hand, turning the indicator through a 180° arc.
- Note the indicator reading.
  The difference between the zero reading at the right-hand side and the reading at the left-hand side is the amount that the head is out of alignment.

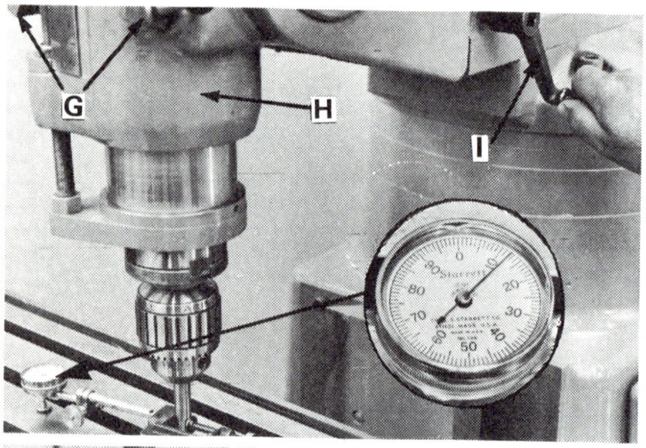

6. • Loosen the lock nuts or screws (G) on the head (H) about one-quarter of a turn.
- With a crank (I) turn the head swivelling shaft until the indicator needle moves half the difference between the two readings.
- Tighten the head lock nuts or screws (G).
- Recheck and readjust the head until the indicator readings are the same at both ends of the table.

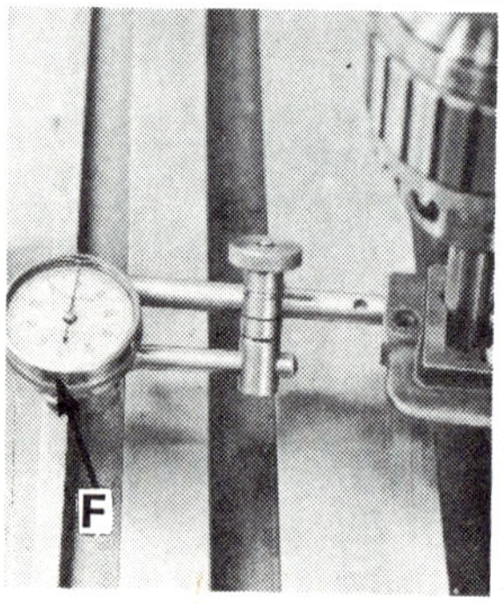

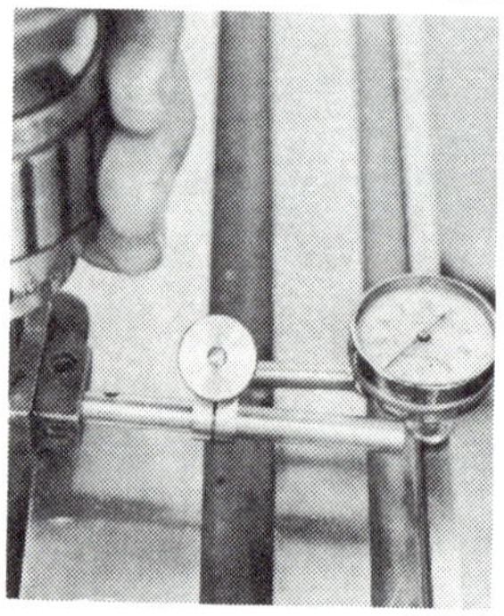

7. • Turn the spindle until the indicator is at the front of the table.
- Raise the table until the indicator registers about one-quarter of a turn.
- Set the indicator bezel (F) to zero.
- Rotate the spindle by hand one-half of a turn (180°).
- The difference between the indicator reading at the front and back edge of the table is the amount the head is out of alignment.

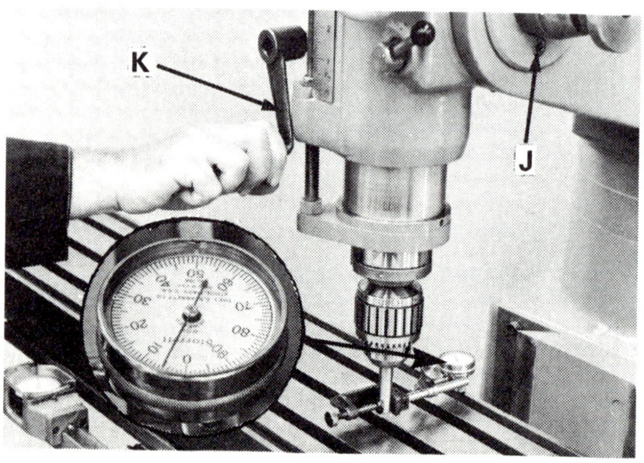

8. • Loosen the head lock screws (J) about one-quarter of a turn.
- With a crank (K) tilt the head until the indicator registers one-half the difference between the two readings.
- Tighten the lock screws (J) and recheck the alignment by turning the spindle 180°.

ALIGNING THE VISE

In order to produce edges which are square and/or parallel to the sides or center line of a workpiece, the vise jaws must be aligned with the table slots.

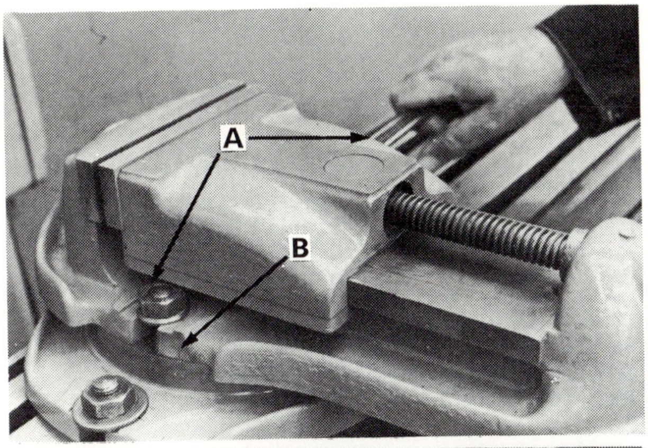

1.
- Loosen both lock nuts (A) about one-half of a turn.
- Swivel the vise so that the index line (B) is in line with the 90° graduation line.
- Tighten the lock nuts (A) *lightly*.
- Remove any burrs from the jaws.
- Clean the vise thoroughly.

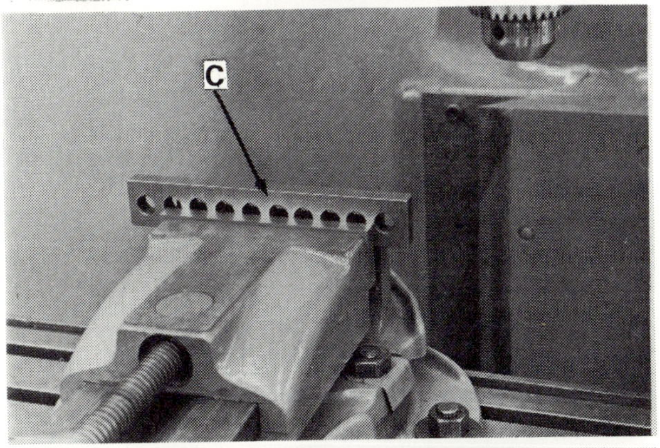

2.
- Select a parallel slightly longer than the vise jaws.
- Remove any burrs with an oilstone and clean it thoroughly.
- Mount the parallel (C) in the vise with about 1" (25.4 mm) above the jaws.

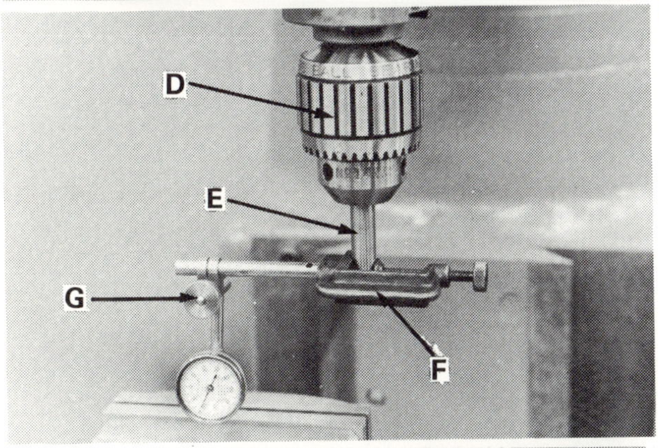

3.
- Mount a drill chuck (D) in the spindle of the machine.
- Fasten a ½" (12.7 mm) diameter rod (E) in the drill chuck.
- Mount the dial indicator assembly (F) on the rod (E).
- Adjust the indicator so the plunger is facing the column of the machine.
- Tighten the indicator clamp nut (G).

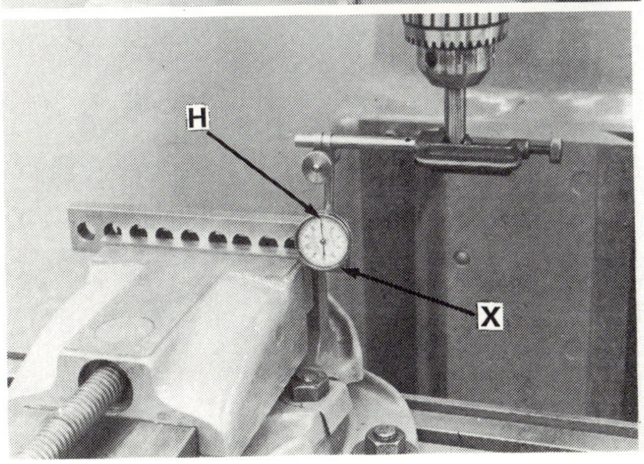

4.
- Adjust the table until the indicator plunger is at the right end (X) of the parallel.
- Adjust the table height until the plunger is about ¼" (6.35 mm) from the top edge of the parallel.
- Move the table until the indicator needle (H) registers ¼ of a revolution.

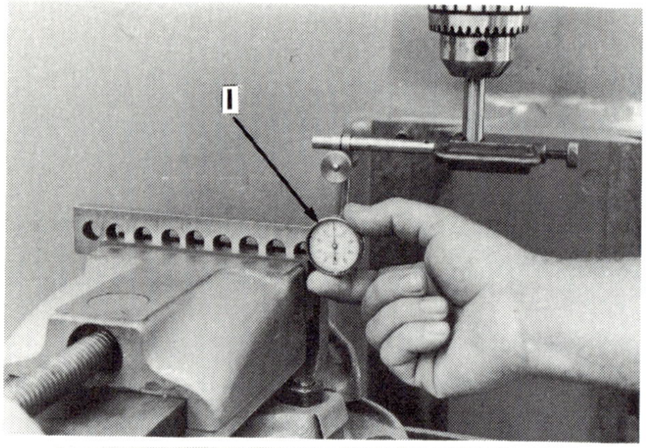

5. • Set the indicator bezel (I) to zero.
CAUTION: Be sure not to rotate the spindle when setting the bezel.

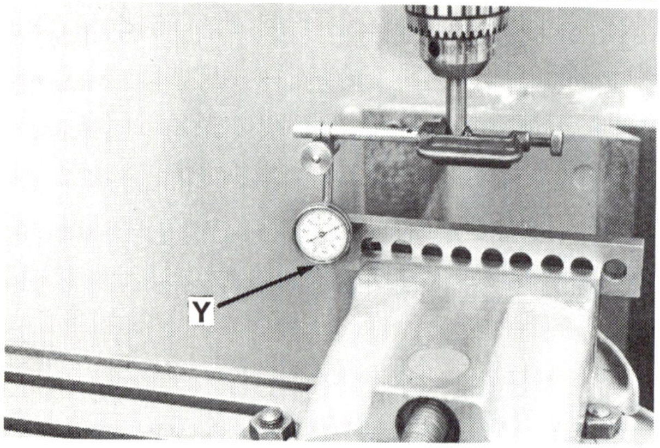

6. • With the table traverse handwheel, move the table lengthwise until the indicator is at the left end (Y) of the parallel.
• Note the indicator reading at this point and compare it to the reading at the other end of the parallel.

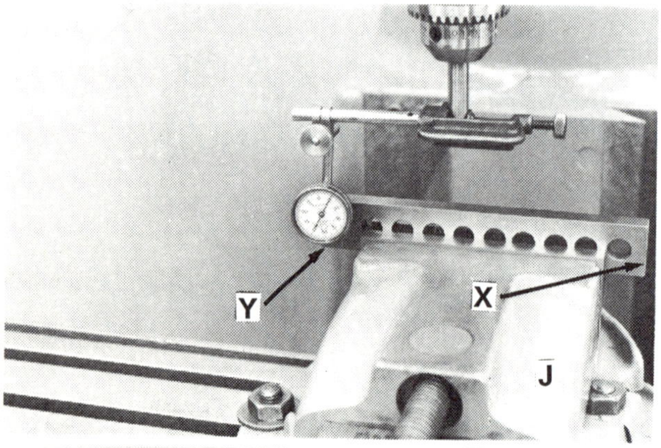

7. • Notice where the indicator reading is higher.
(a) If the reading is higher at (Y), tap the vise clockwise at (J) with a soft-faced hammer one-half the difference between the two indicator readings.
(b) If the reading is higher at (X), tap the vise counterclockwise.
NOTE: Always tap the vise so that the parallel moves *away from* and not towards the indicator.

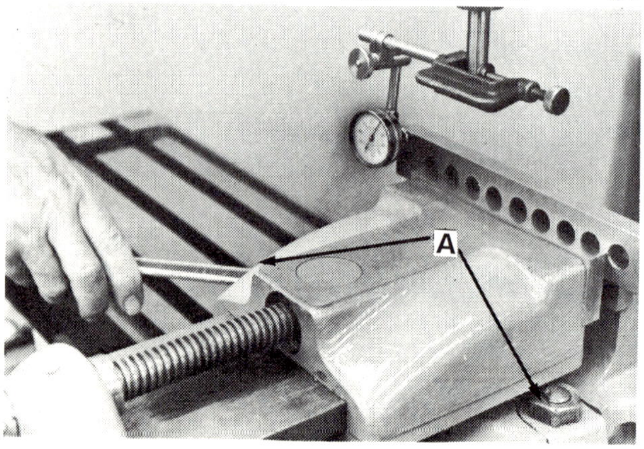

8. • Tighten the lock nuts (A) securely.
• Set the indicator bezel to zero.
• Move the table until the indicator is at the other end of the parallel.
• Check for any difference in the indicator reading between both ends.
• If necessary, repeat the procedure in Frame 7 until the indicator readings at both ends are the same.

To obtain a flat surface on work, the vertical head should be aligned at 90° with the surface of the table. The cutting tool used may be either an end mill or a flycutter.

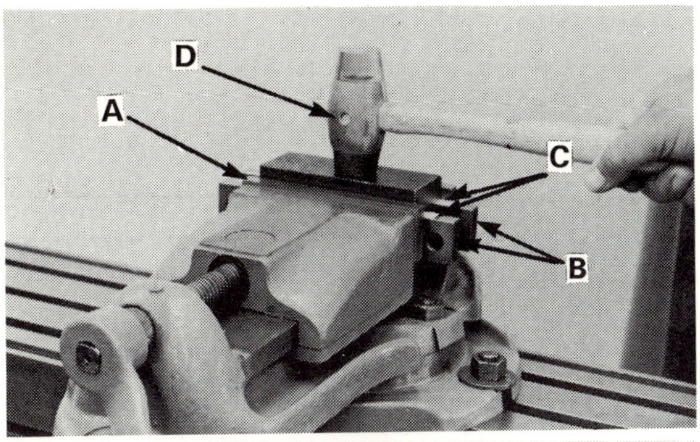

1. • Swivel the vise so that the jaws (A) are parallel with the table slots.
   • Remove all burrs and clean the vise and work.
   • Set the work, in the center of the vise, on parallels (B) and paper feelers (C).
   The work should be about ¼" (6.35 mm) above the vise jaws.
   • Tighten the vise securely.
   • Tap the work down with a soft-faced hammer (D) until all paper feelers are tight.

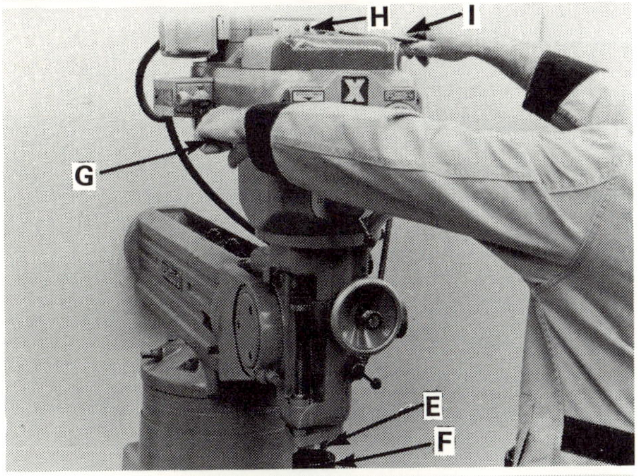

2. • Shut off the power to the machine.
   • Clean the taper in the spindle (E).
   • Place the flycutter (F) into the spindle.
   • Pull the spindle brake handle (G) forward with one hand.
   • Tighten the draw-bolt (H) securely with a wrench (I).

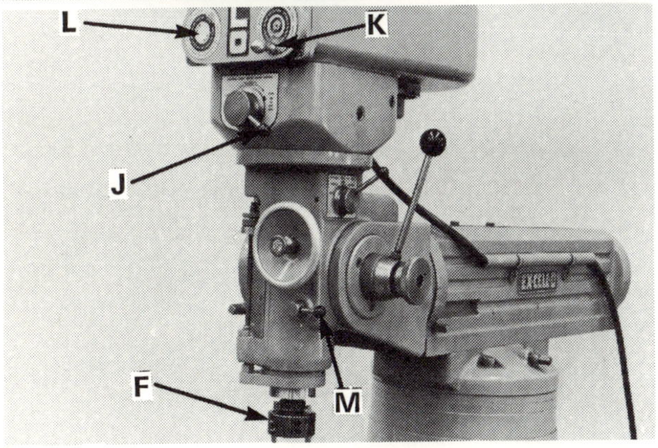

3. • Calculate the spindle speed for the material being cut.

$$\text{rpm} = \frac{CS \times 4}{D}$$

   • Move lever (J) to the high or low range as required.
   • Start the machine.
   • Turn the spindle speed selector (K) until the proper speed is indicated on the spindle speed indicator (L).
   • Check that the flycutter (F) is revolving in the proper direction.

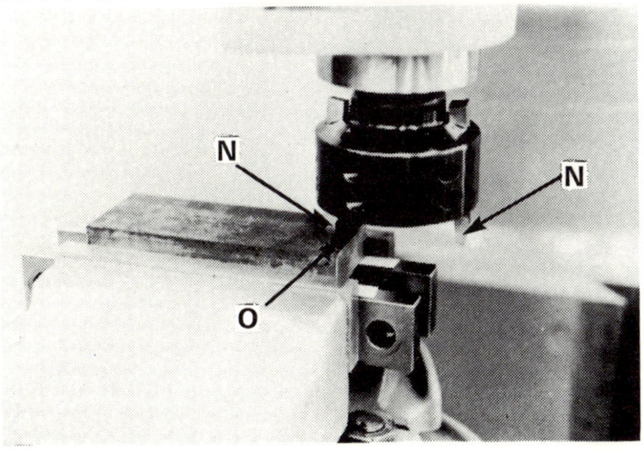

4. • Raise the quill to its upper position and tighten the quill lock lever (M) (see Frame 3).
   • Adjust the table to align the center line of the work with the center of the cutter.
   • Move the table until the cutter is over the right-hand end of the work.
   • Raise the table until the cutting tool (N) just touches the work surface (O).
   • Move the table so that the work clears the cutter.

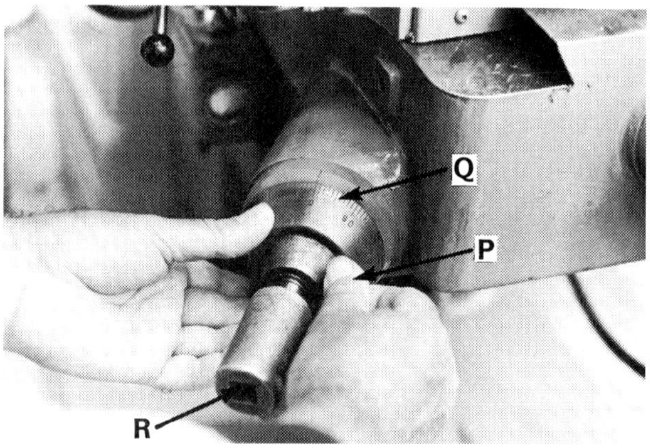

5. • Loosen the lock screw (P) on the graduated collar (Q).
   • Set the graduated collar to zero. DO NOT turn the elevating screw (R).
   • Tighten the graduated collar lock screw.

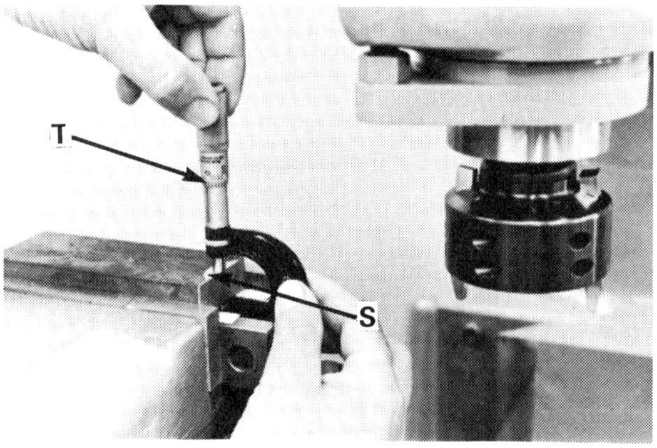

6. • Set a .030 (.76 mm) depth cut by turning the elevating screw handle clockwise.
   NOTE: Always lock the knee clamp before cutting; unlock it before setting the depth of cut.
   • Start the machine and take a trial cut (S) for about ¼" (6.35 mm) long.
   • Stop the machine and measure the thickness of the work with a micrometer (T).

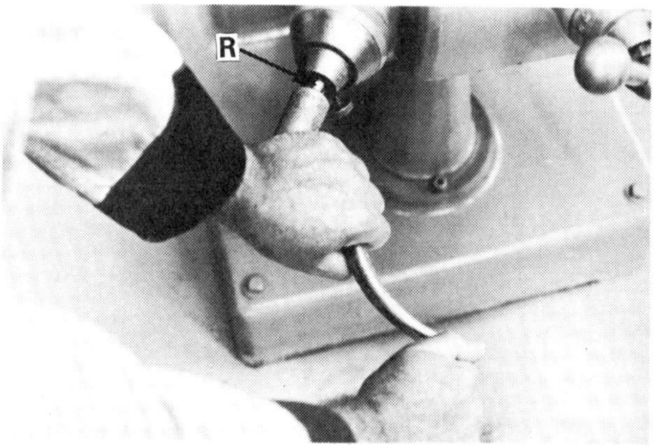

7. • Set the depth of cut to within .030 (.76 mm) of the finished size by turning the elevating screw (R) clockwise.
   • Take the roughing cut by feeding the work automatically or by hand at a steady rate under the revolving cutter.

8. • Stop the machine and measure the work.
   • Raise the table the required amount for the finished size of the work.
   • Take a trial cut ¼" (6.35 mm) long and then shut off the machine.
   • Measure the work and reset the depth of cut if necessary.
   • Take the finish cut across the work.
   • Shut off the machine and remove the work from the vise.

MACHINING WORK SQUARE AND PARALLEL

In order to machine the sides of a piece of work square and parallel, it is important that each side be machined in a definite order. To machine work square and parallel, the toolhead must be at right angles (90°) to the surface of the table.

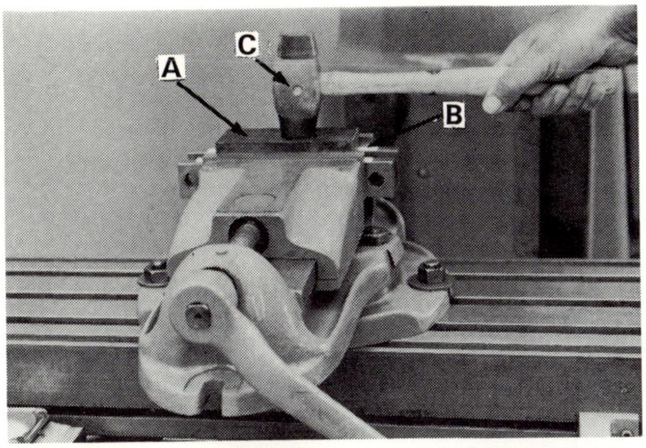

1. - Remove the burrs and clean the vise and parallels.
   - Set the work (A) with its largest side up, in the center of the vise.
   - Place a short paper feeler (B) between each corner of the work and the parallels.
   - Tighten the vise securely.
   - Tap the work down with a soft-faced hammer (C) until all paper feelers (B) are tight.

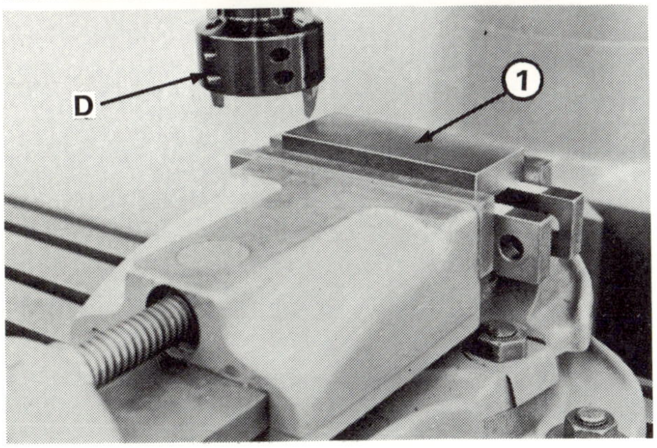

2. MACHINING SIDE #1
   - Mount a flycutter (D) in the spindle.
   - Set the spindle speed $\left(\text{rpm} = \dfrac{\text{CS} \times 4}{\text{Dia.}}\right)$
   - Raise the table until the revolving cutter just touches near the right-hand end of side #1.
   - Set the elevating screw collar to zero.
   - Move the work clear of the cutter.
   - Set a .030 (.76 mm) depth of cut.
   - Using a steady feed rate, machine side #1.

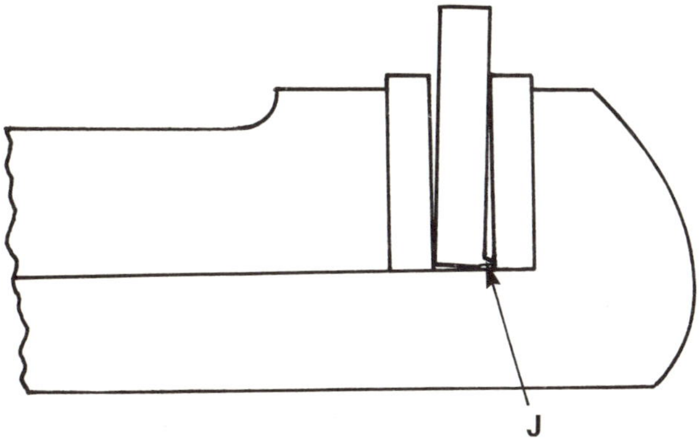

3. - Remove the work from the vise.
   - Remove the burrs from the edges of the surface just machined.
   - If the burrs are not removed such as at (J), they will prevent the machined surface from being held flat against the solid jaw.
   - If the work is tilted and not held flat against the solid jaw of the vise and side #2 is machined, it will not be square (at 90°) to side #1.

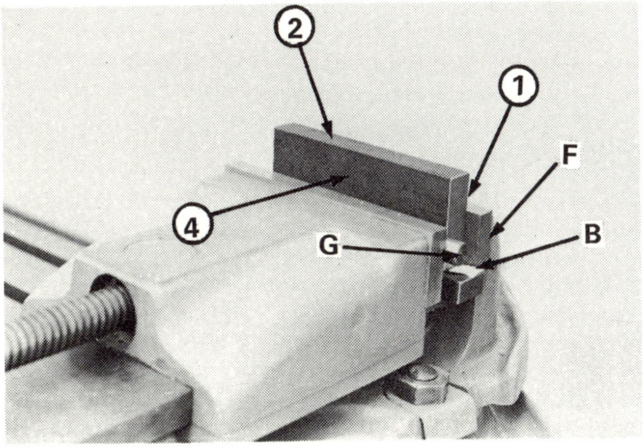

4. MACHINING SIDE #2
   - Clean the vise and work thoroughly.
   - Place side #1 against the solid jaw (F) and side #2 up. Use a parallel if required.
   - Place a round bar (G) between the movable jaw and side #4.

     NOTE: The round bar must be in the center of the amount of work being *held inside the vise jaws.*
   - Tighten the vise securely and tap the work down until the paper feelers (B) are tight.

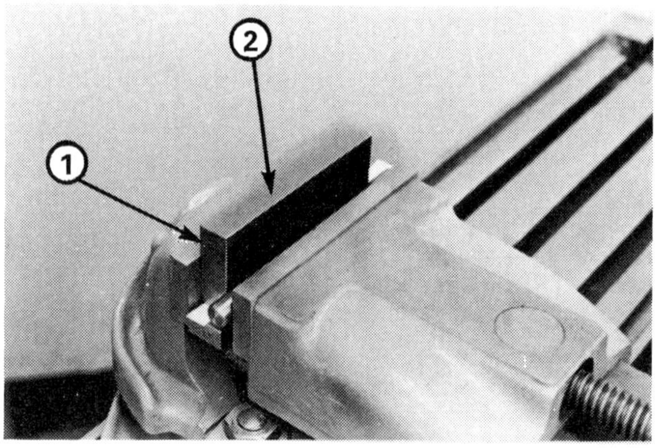

5. • Raise the table until the revolving cutter just touches the right-hand end of side #2.
   • Move the table until the work clears the cutter.
   • Set a .030 (.76 mm) depth of cut using the vertical screw graduated collar.
   • Using a steady feed rate, machine side #2.
   • Remove all burrs from the edges of the work.
   • Use a square to check that side #1 is square (at 90°) to side #2.

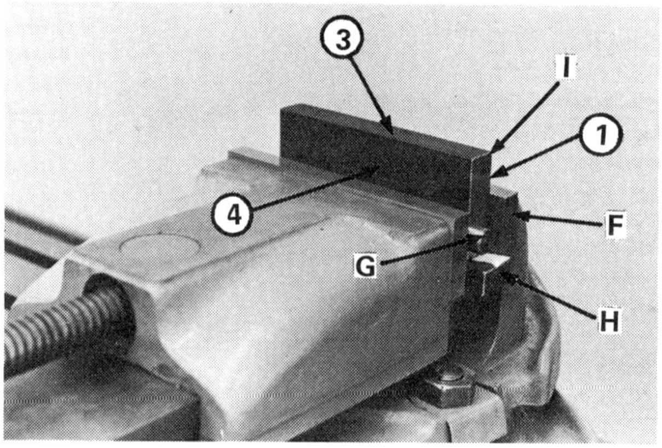

6. MACHINING SIDE #3
   • Clean the vise and work thoroughly.
   • Place side #1 against the solid jaw (F) with side #2 resting on the parallel (H).
   • Place a round bar (G) between side #4 and the movable jaw.
   • Tighten the vise securely.
   • Push the parallel to the left until it is ¼" (6.35 mm) to the left of the edge of the work (I).
   • Tap the work down to seat it on the parallel.

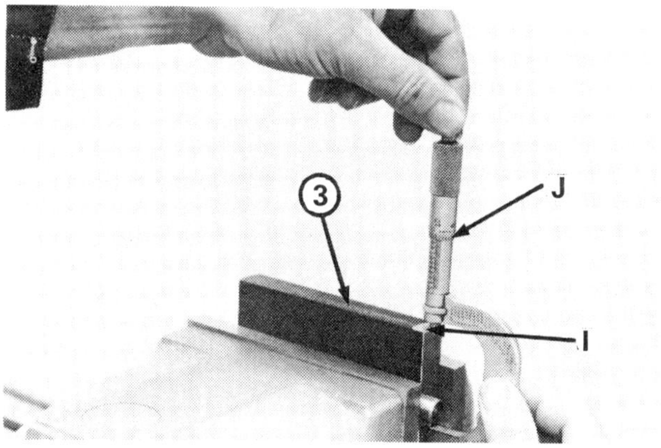

7. • Start the machine.
   • Raise the table until the revolving cutter just touches side #3 near the right-hand end of the work (I).
   • Move the table until the cutter clears the right-hand end of the work.
   • Raise the table .010 (.25 mm) and take a trial cut for about ¼" (6.35 mm) from side #3.
   • Stop the machine and measure the width of the work with a micrometer (J).

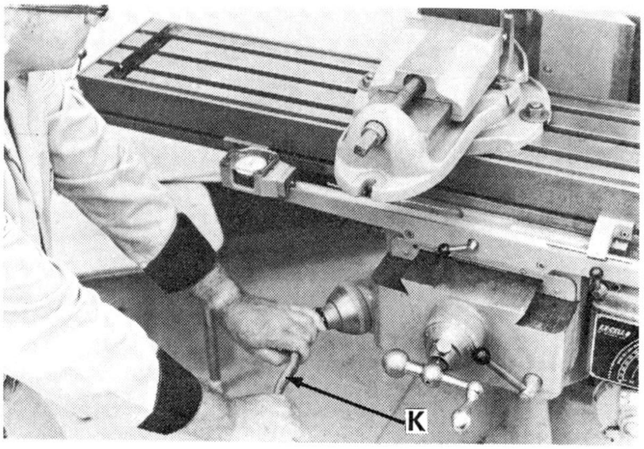

8. • Calculate the difference between the finish size and the trial cut size.
   • With the elevating screw crank (K) raise the table the required amount.
   • Start the machine and take a trial cut ¼" (6.35 mm) long.
   • Stop the machine and measure the trial cut size.
   • Readjust the table if necessary and then machine side #3.

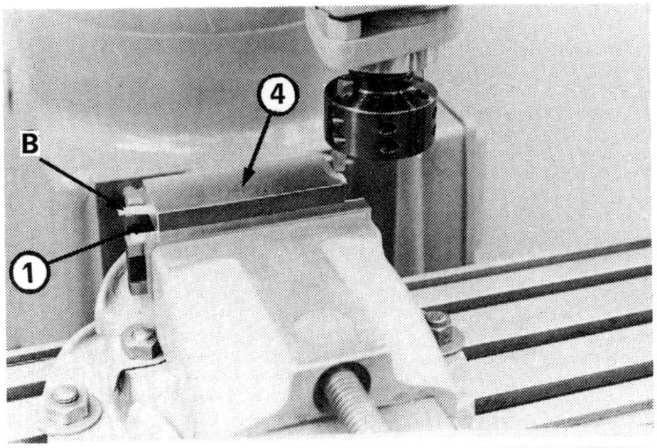

9. MACHINING SIDE #4
   - Place side #1 on parallels using paper feelers (B) under each corner of the work.
   - Tighten the vise securely and tap the work down until all feelers are tight.
   - Take a light trial cut at the end of side #4.
   - Measure the size of the trial cut with a micrometer.
   - Raise the table the correct amount.
   - Machine side #4 to size.

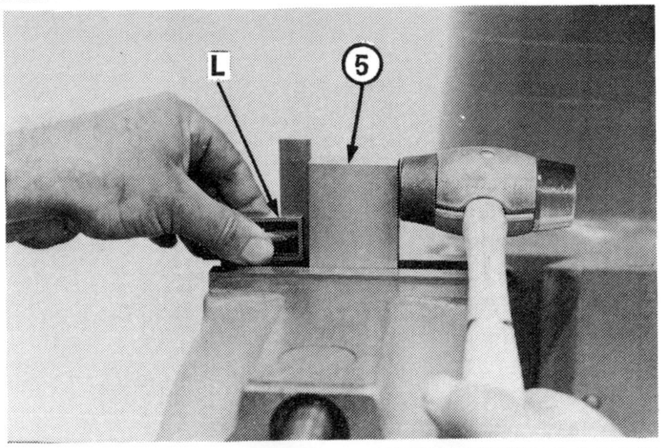

10. MACHINING THE ENDS
    - Set the work in the center of the vise with side #5 up and tighten the vise lightly.
    - Hold a square (L) down firmly on the top of the vise jaws.
    - Use light side pressure to bring the blade of the square against the work.
    - Tap the work until its edge is aligned with the blade of the square.
    - Tighten the vise securely.
    - Machine side #5 until the surface cleans up.

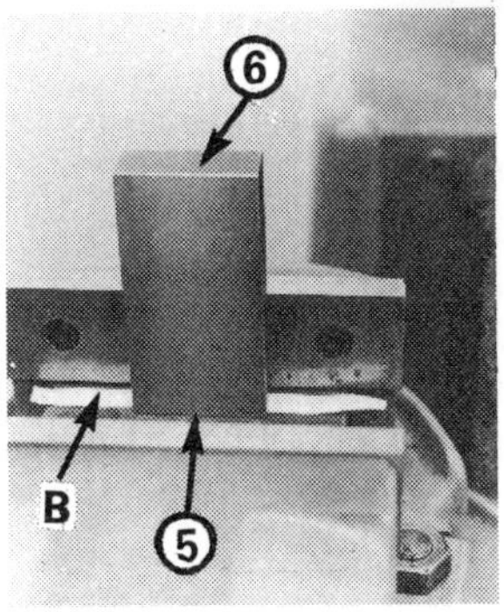

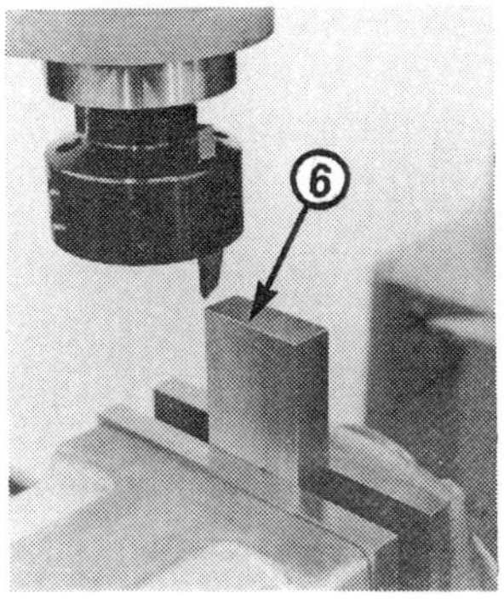

11. - Remove the burrs from the edges of side #5.
    - Clean the vise thoroughly.
    - Place side #5 on the base of the vise with a paper feeler under each end.
    - Tighten the vise securely.
    - Tap the work down until the paper feelers (B) are tight.
    - Take a cut off side #6 until the surface cleans up.

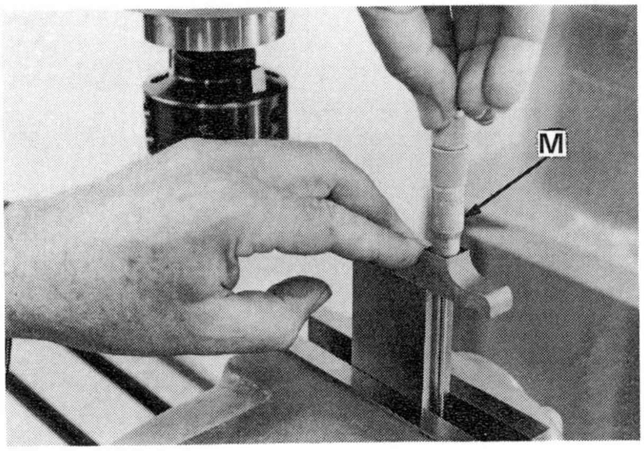

12. - Remove all burrs from side #6.
    - Clean all cuttings from the work and the vise.
    - Measure the height of the workpiece using a depth micrometer (M).
    - Raise the table the required amount and machine side #6.
    - Shut off the machine and remove the burrs from side #6.
    - Remove the work from the vise.
    - Check the surface for squareness.

To machine the ends of the work square with the top and the side, the toolhead and the vise must be aligned with the table travel. The ends on most flat work can be machined by using an end mill or a flycutter.

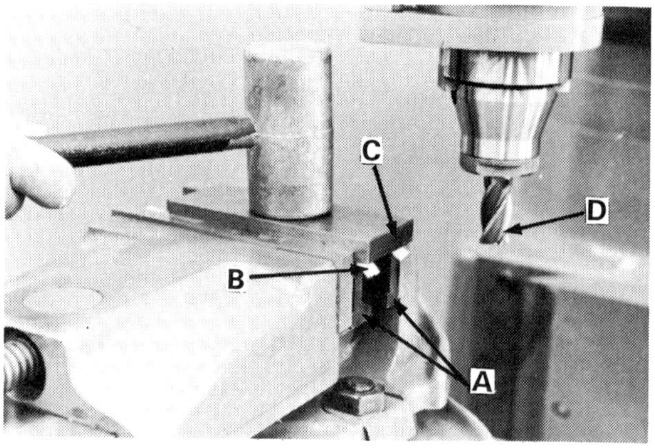

1. • Align the vise jaws parallel with the table slots.
   • Remove all burrs and clean the work, parallels and vise.
   • Set the work on parallels (A) and paper feelers (B).
   • Position the work so that the end (C) projects past the vise jaws and parallels by about ¼" (6.35 mm).
   • Tighten the vise securely and tap the work down until all paper feelers are tight.
   • Mount an end mill (D) in the spindle.

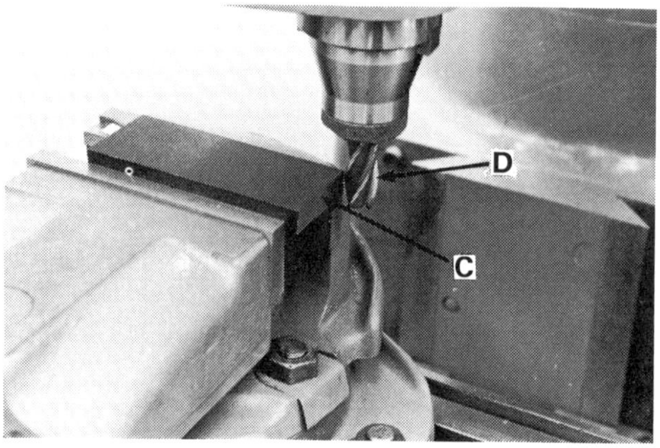

2. • Set the spindle to the correct speed.
   • Carefully move the table until the revolving end mill (D) touches the surface (C).
   • Move the work clear of the cutter using the crossfeed handle.
   • Set a .030 (.79 mm) depth of cut using the graduated collar on the table traverse handle.
   • Lock the knee and the table traverse.
   • Machine the surface (C) by feeding the work across the cutter with the crossfeed handle.

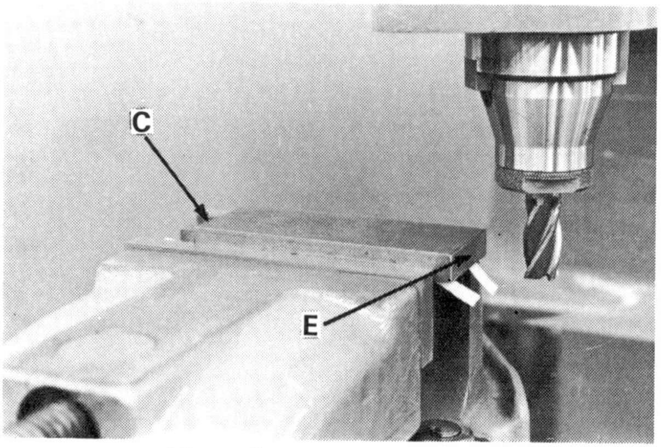

3. • Loosen the table traverse lock and move the work clear of the cutter.
   • Remove the workpiece from the vise.
   • Remove the burrs from the end (C).
   • Clean the vise and parallels.
   • Reverse the workpiece so that end (E) projects about ¼" (6.35 mm) past the vise jaws and the parallels.
   • Tighten the vise securely and tap the work down until all paper feelers are tight.

4. • Start the spindle revolving.
   • Move the table until the end mill just touches the work surface (E).
   • Tighten the table traverse lock.
   • Take a cleanup cut off end (E) by either climb or conventional milling. (See Step 6 HORIZONTAL MILLING OPERATION #4.)
   • Stop the spindle.
   • Move the work clear of the cutter using the crossfeed handle.

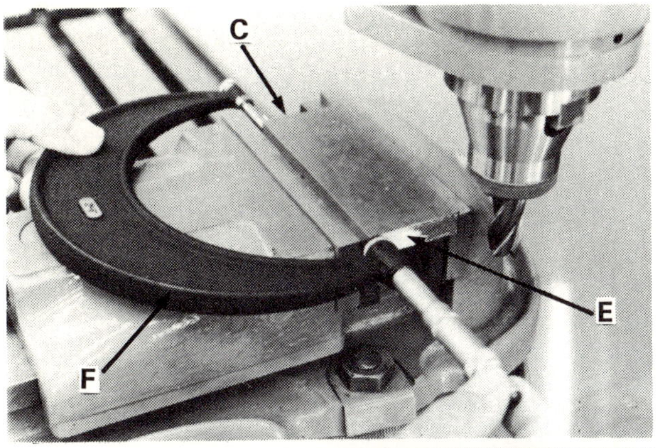

5. • Clean edges (C) and (E).
   • Measure the length of the work with a micrometer (F) or a vernier caliper.
   • Calculate the amount to be removed.
   • Set the depth for each cut by using the table traverse screw graduated collar.
   • Tighten the table traverse lock and machine the work to the correct length.

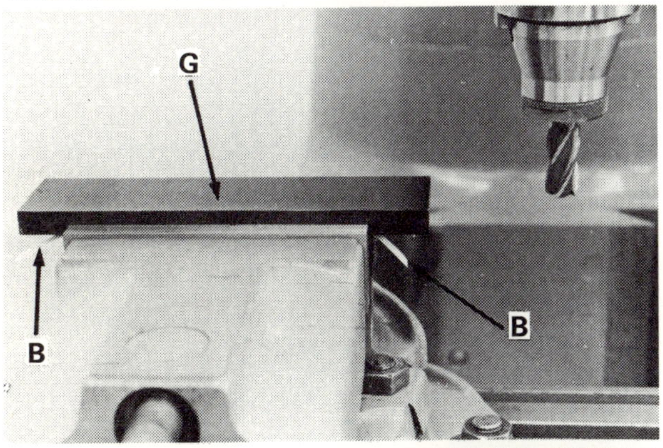

6. MACHINING ENDS OF LONG WORK
   • Remove all burrs and clean the vise and parallels.
   • Set the work on the parallels and paper feelers (B).
   • Center the workpiece (G) in the vise.
   • Tighten the vise securely and tap the work down until all paper feelers are tight.

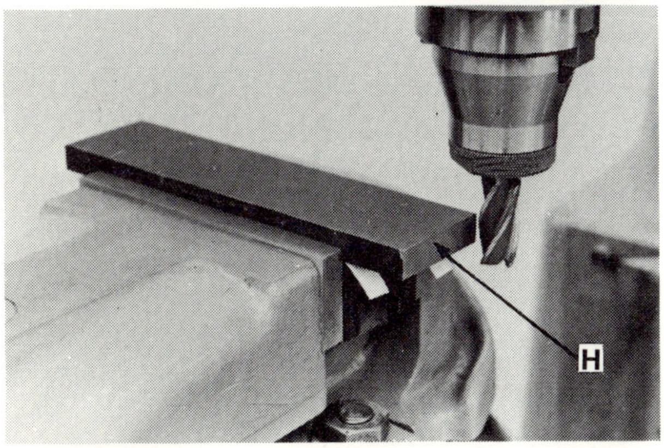

7. • Start the spindle and touch the end (H) of the work to the cutter using the table traverse handle.
   • Move the work clear of the cutter with the crossfeed handle.
   • Using the table traverse graduated collar, set a depth of cut of about .030 (.76 mm).
   • Tighten the table traverse lock.
   • Machine the end (H) using the crossfeed handle.

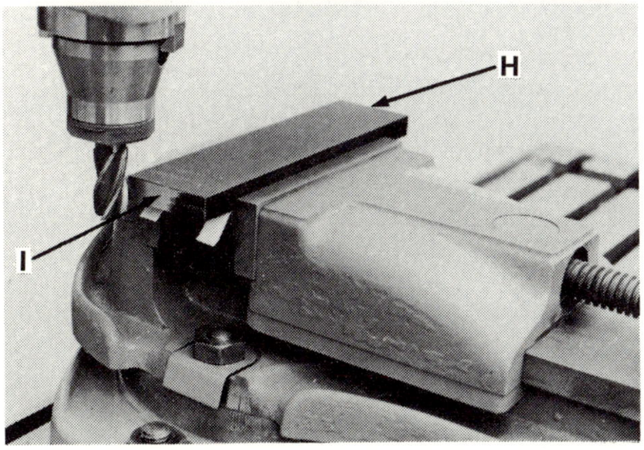

8. • Move the table so that the cutter touches the edge (I) of the work.
   • Take a cleanup cut from this edge.
   • Move the work clear of the cutter using the crossfeed handle.
   • Measure the work and calculate the amount to be removed.
   • Set the depth of cut using the table traverse graduated collar.
   • Engage the table lock clamp.
   • Machine the workpiece to size.

Angular surfaces can be produced on a milling machine by swivelling the toolhead to the required angle or by setting the work on an angle in the vise.

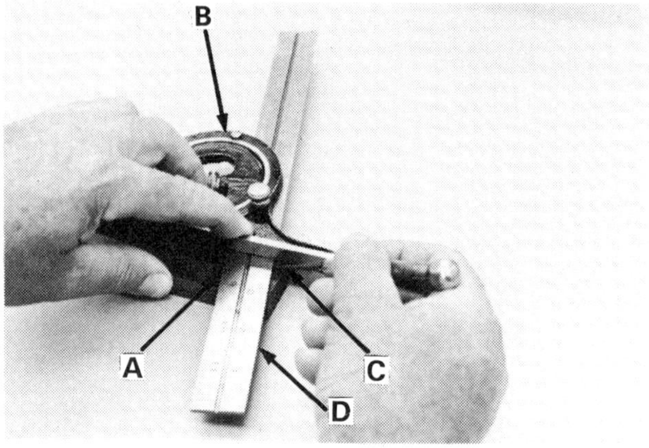

1. **SWIVELLING THE TOOLHEAD**
   - Remove all burrs from the workpiece.
   - Apply layout dye to the workpiece (A).
   - Set the protractor (B) to the required angle.
   - Hold the base of the protractor against the side of the workpiece.
   - With a sharp scriber (C), held at an angle against the edge of the rule (D), mark a line on the workpiece.

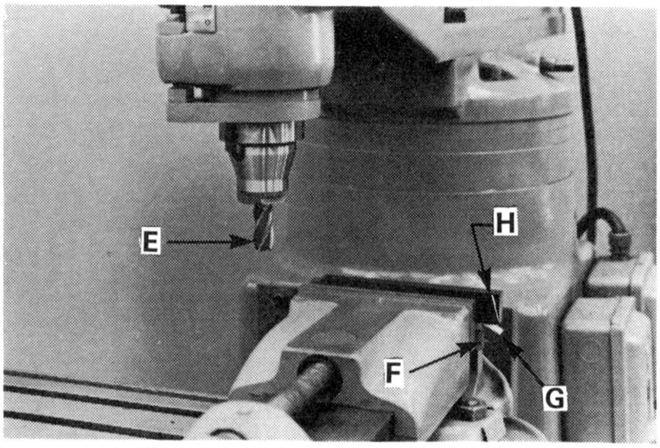

2.
   - Mount an end mill (E) whose cutting edge is slightly longer than the angular surface to be cut.
   - Align the vise jaws parallel with the table slots.
   - Set the work on parallels (F) and paper feelers (G).
   - Move the work so that the layout line (H) extends about ½″ (12.7 mm) beyond the end of the vise jaws and the parallels.
   - Tighten the vise securely and tap the work down until all paper feelers are tight.

3.
   - Loosen the toolhead lock screws about ¼ of a turn.
   - Turn the swivelling shaft (I) and set the toolhead to the required angle as indicated by the graduated lines (J).
   - Tighten the toolhead lock screws securely.

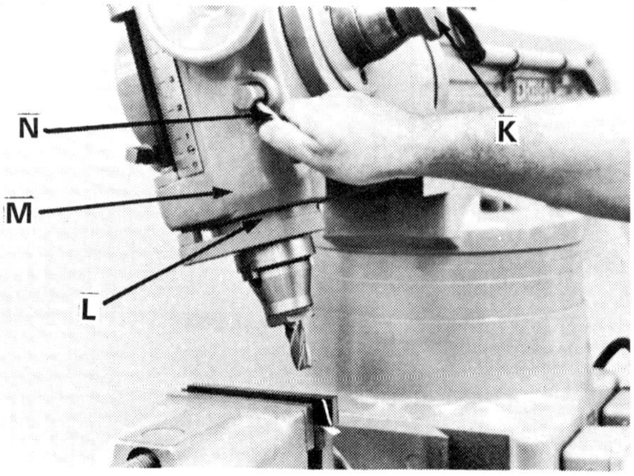

4.
   - With the quill-feed hand lever (K) move the quill (L) up as far as possible into the housing (M).
   - Tighten the quill lock (N).

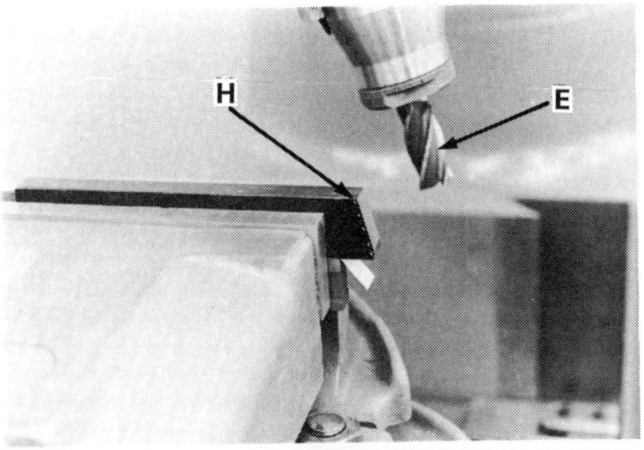

5. • Raise the table until the end of the cutter (E) clears the bottom edge of the work.
   • Start the spindle.
   • Set the depth of cut with the table traverse handle.
   • Tighten the table traverse lock.
   • Rough cut the work to within 1/32" (.79 mm) of the layout line (H) using the crossfeed handle.
   • Stop the machine and move the work away from the cutter using the crossfeed handle.

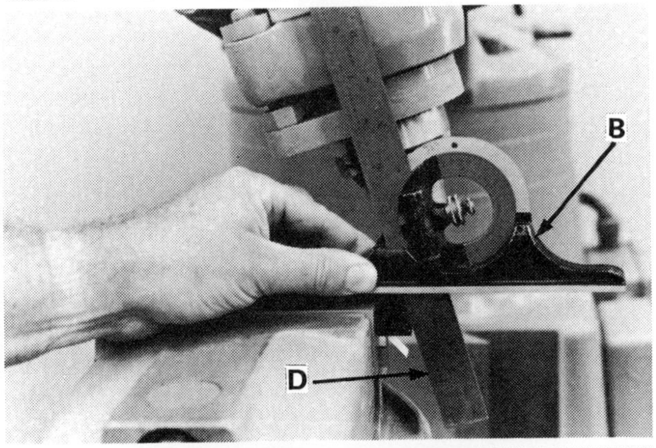

6. • Remove the burrs from the top edge of the work.
   • Place the base of the protractor (B) against the top surface of the work.
   • Slide the protractor along the work until the edge of the rule (D) contacts the machined end of the work.
   • Check the angle of the machined surface.
   • Adjust the toolhead if the angle is not correct.

7. • Set the depth of the finish cut with the table traverse handle.
   • Tighten the table traverse lock.
   • Take the finish cut using the crossfeed handle.
   • The work is correct when the layout line (H) and the punch marks are cut in half.

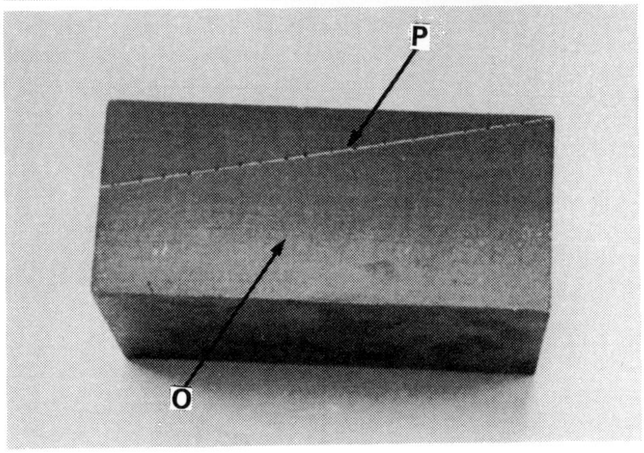

8. SETTING THE WORK ON AN ANGLE
   • Remove all the burrs and clean the workpiece.
   • Coat the surface (O) with layout dye.
   • Layout the required angle using a protractor and a scriber.
   • Space light prick punch marks on the layout line (P) about 3/8" (9.52 mm) apart.

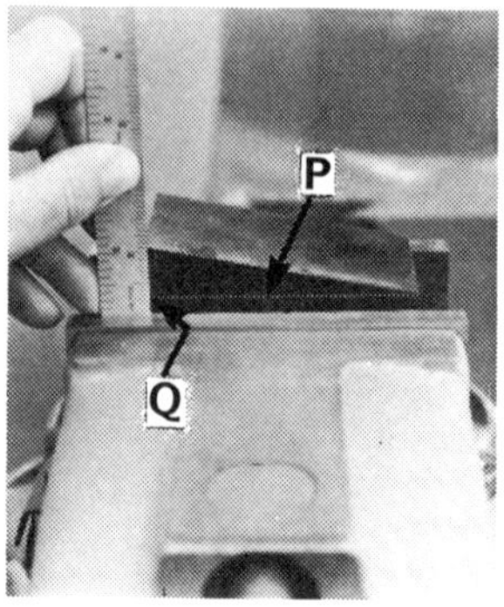

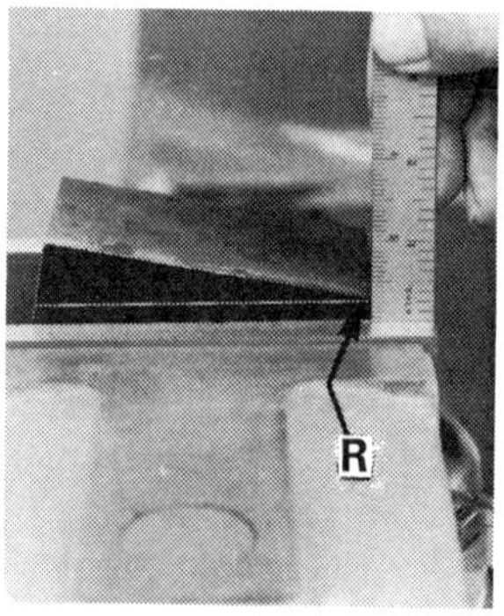

9. ALIGNING USING A RULE
   - Clean the vise thoroughly.
   - Clamp the work in the center of the vise with the layout line (P) about ¼" (6.35 mm) above the vise jaws.
   - Measure the distance from the vise jaw to the layout line at (Q) and (R).
   - Tap the proper end of the work until the height of the layout line is the same as (Q) and (R).
   - Tighten the vise securely.

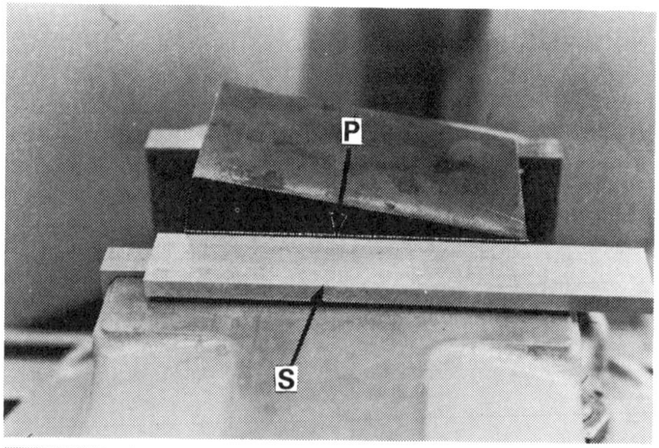

10. ALIGNING USING A PARALLEL
    - Clean the vise thoroughly.
    - Clamp the work in the center of the vise with the layout line (P) about ¼" (6.35 mm) above the vise jaws.
    - Place a parallel (S) on the top of the vise jaws.
    - Tap the work down until the layout line (P) is even with the parallel (S).
    - Tighten the vise securely.

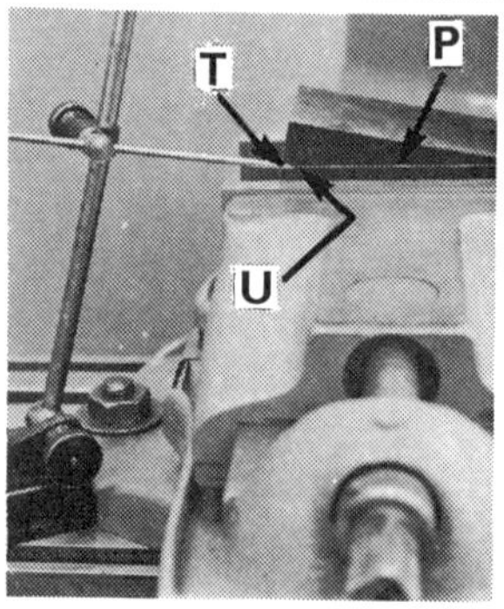

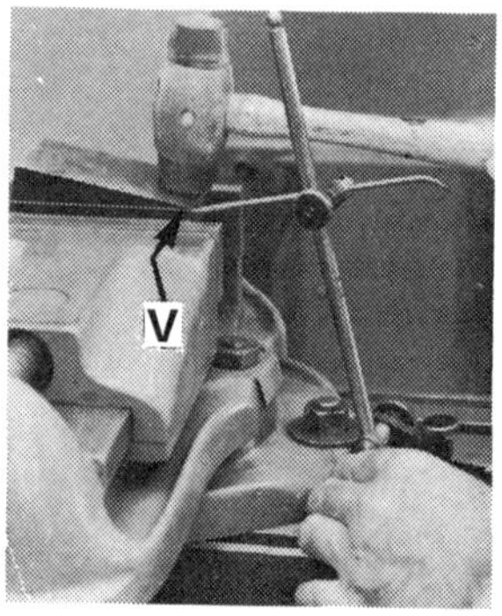

11. ALIGNING USING A SURFACE GAGE
    - Mount the work in the center of the vise with the layout line (P) about ¼" (6.35 mm) above the vise jaws.
    - Set the surface gage scriber point (T) to the layout line at the end (U).
    - Move the surface gage to the end (V).
    - Tap the work down until the line at both ends is even with the scriber point (T).
    - Tighten the vise securely.

12.
    - Mount a flycutter (W) in the spindle.
    - Check the rotation of the cutter.
    - Set the proper spindle speed.
    - Take successive roughing cuts to within $\frac{1}{32}$" (.79 mm) of the layout line (P).
    - Raise the table until the cut just splits the layout line.
    - Take a finish cut across the surface.
    - Remove the work from the vise and file off the burrs.

MOUNTING AND REMOVING GRINDING WHEELS

There are generally two types of grinder spindles. On one type, a collar and nut must be removed to change the wheel. The other type uses an adaptor to hold the wheel and has provision for balancing the wheel. Before any wheel is mounted, it should be tested for cracks.

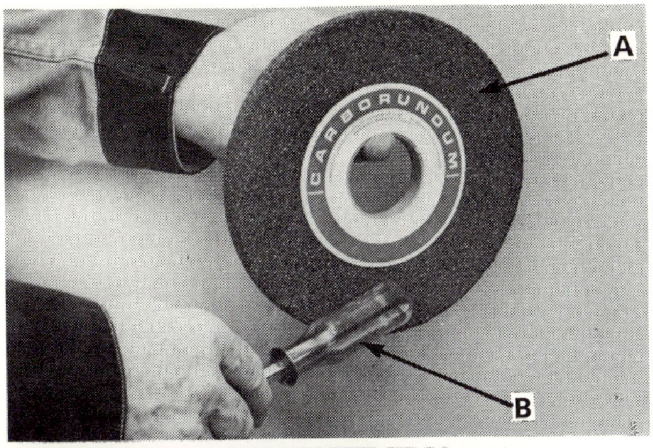

1. • Support the grinding wheel (A) on a finger or with a small pin through the hole of the wheel.
   • Lightly tap the wheel at four points about 90° apart with a plastic or wooden handled screwdriver (B).
   • A good vitirfied bonded wheel will produce a clear metallic ring while a cracked wheel will not produce a ring.

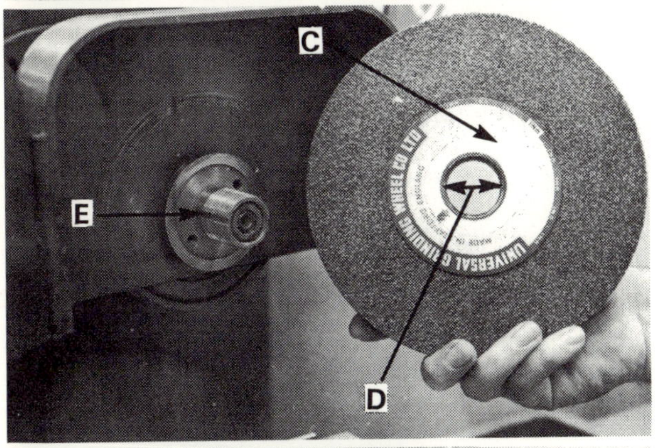

2. MOUNTING A WHEEL ON A STRAIGHT SPINDLE
   • Inspect the grinding wheel for chips and nicks.
   • Examine the blotter (C) on each side of the grinding wheel and replace if not in good condition.
   • Clean the bore (D) of the wheel.
   • Clean the collar and the spindle (E).

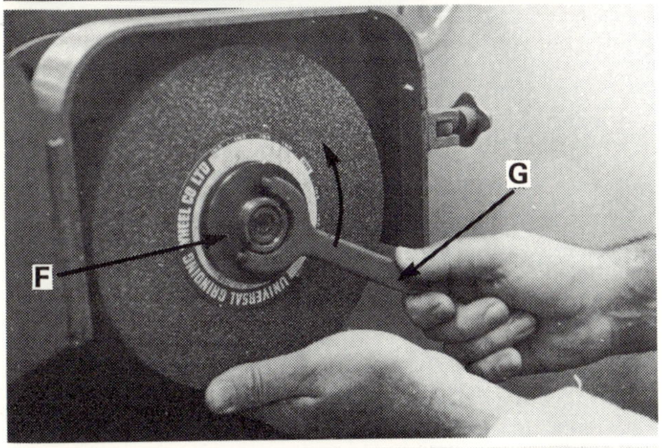

3. • Place the wheel on the spindle.
   NOTE: The wheel should slide onto the spindle freely and *never* be forced on.
   • Thread the nut (F) counterclockwise onto the spindle by hand.
   • Grasp the lower surface of the wheel.
   • Tighten the nut securely with the proper wrench (G).

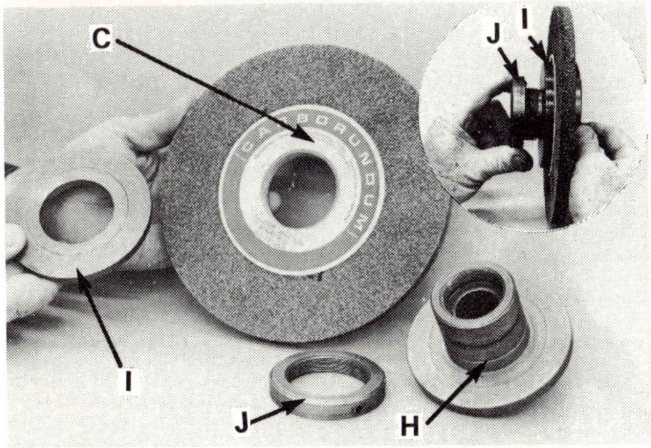

4. MOUNTING A WHEEL ON A TAPERED ADAPTOR
   • Inspect the wheel and see that the blotters (C) are in good condition.
   • Clean the adaptor (H), the outer flange (I), and the hole in the wheel thoroughly.
   • Mount the wheel on the adaptor body—it should be a sliding and *not a tight fit*.
   • Mount the outer flange (I) on the adaptor body.
   • Tighten the adaptor nut (J) with the proper wrench.
   • True the wheel. (See Operation #2).

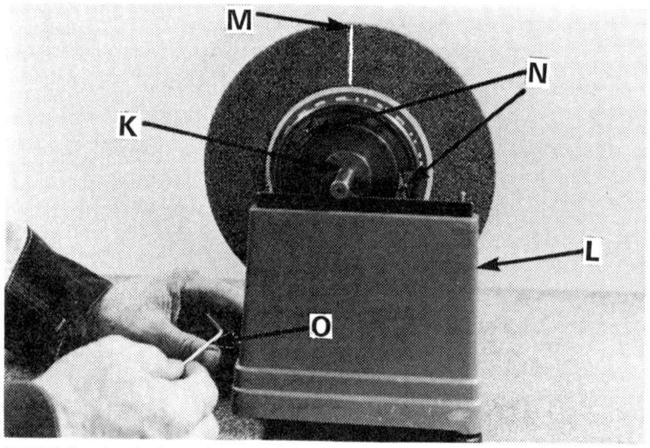

5. BALANCING A WHEEL
   - Clean the tapered hole in the wheel adaptor body.
   - Insert a balancing arbor (K) into the tapered hole in the adaptor.
   - Place the arbor on a balancing stand (L) which has been levelled.
   - When the wheel comes to rest, mark the light side (top) with a chalk mark (M).
   - Adjust the balancing weights (N) with the wrench (O) until the wheel will remain stationary when set at four positions 90° apart.

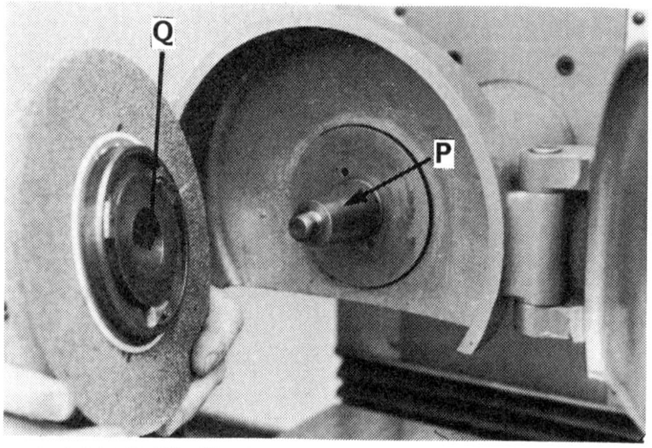

6. MOUNTING THE ADAPTOR AND WHEEL
   - Clean the spindle (P) and the hole (Q) in the wheel adaptor.
   - Mount the adaptor on the spindle.
   - Thread the spindle nut onto the end of the spindle. This is generally a left-hand thread.

7. - Grasp the wheel with one hand.
   - Insert the wrench (R) in the holes in the spindle nut (S).
   - Tighten the nut securely.
   - Turn the grinding wheel *by hand* to check that it clears the wheel housing.
   - Close and lock the housing door (T).

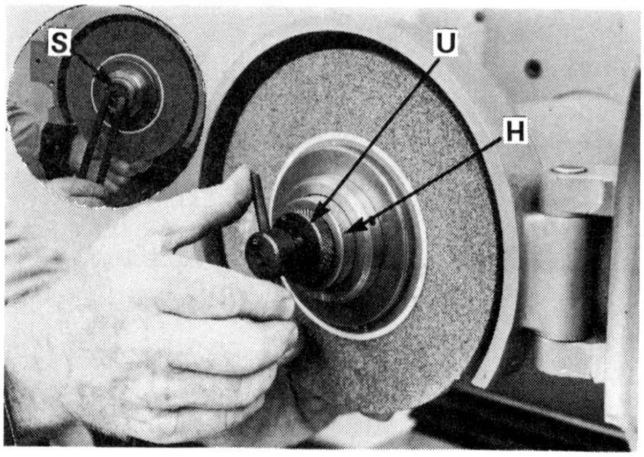

8. REMOVING A GRINDING WHEEL
   - Remove the spindle nut (S).
   - Thread a wheel puller (U) onto the adaptor (H).
   - Grasp the grinding wheel and turn the handle clockwise until the adaptor is loosened on the spindle.
   - Remove the wheel from the spindle.

TRUING AND DRESSING A GRINDING WHEEL

In order to produce accurate work and a good surface finish, the grinding wheel should be trued and dressed at the first sign of inaccurate or poorly finished work.

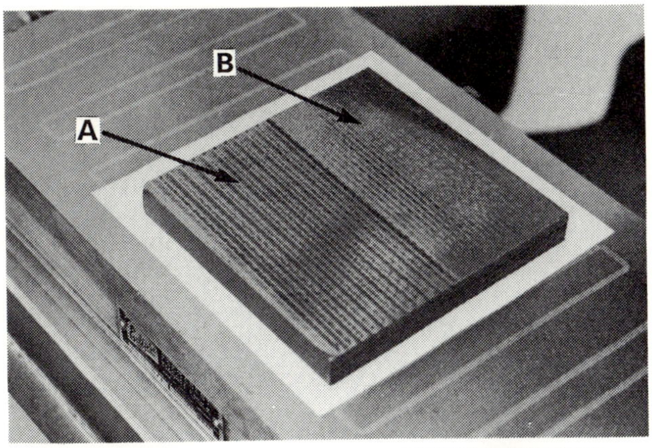

1. • The burned finish and feed lines at (A) and the scratches on the work surface at (B) indicate that the grinding wheel needs dressing.
   • The wheel should also be dressed if irregular patterns, chatter or wavy marks are produced.

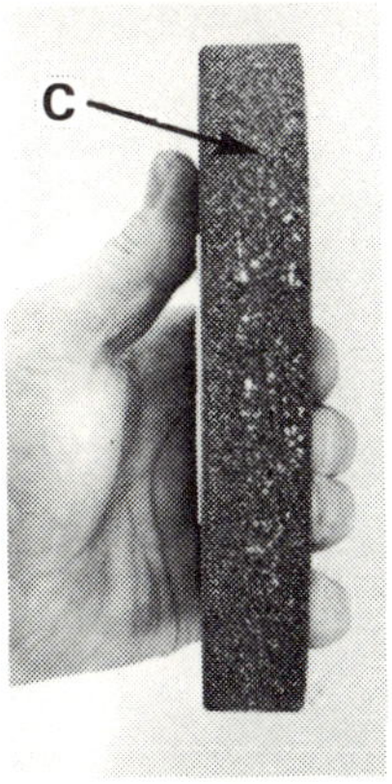

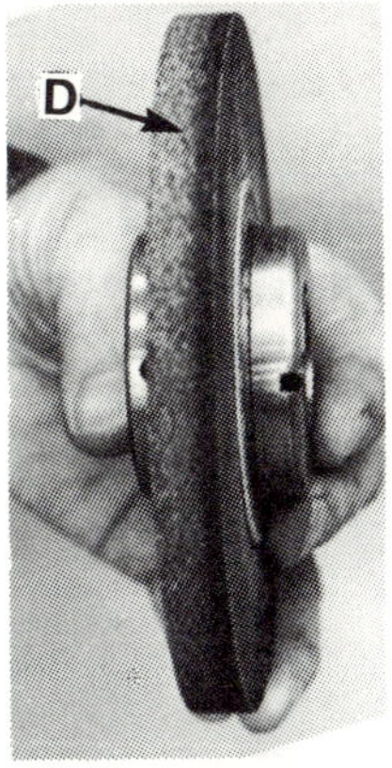

2. • A loaded wheel (C) will produce a poor surface finish.
   A loaded wheel (metal particles imbedded in the wheelface) will not cut properly and will produce burn marks.
   • A glazed wheel (D) will produce a burned or discolored surface on the work.
   • Loaded or glazed wheels should be dressed before using.

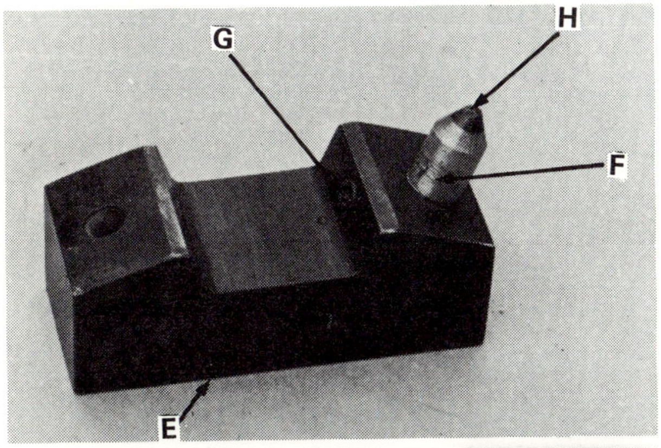

3. • A diamond dresser used for dressing the grinding wheel consists of a flat base (E).
   • A diamond holder (F) mounted in the base at an angle of 10° to 15°, is held securely by means of a set screw (G).
   • The diamond holder should occasionally be rotated in the base in order to keep the diamond (H) sharp.

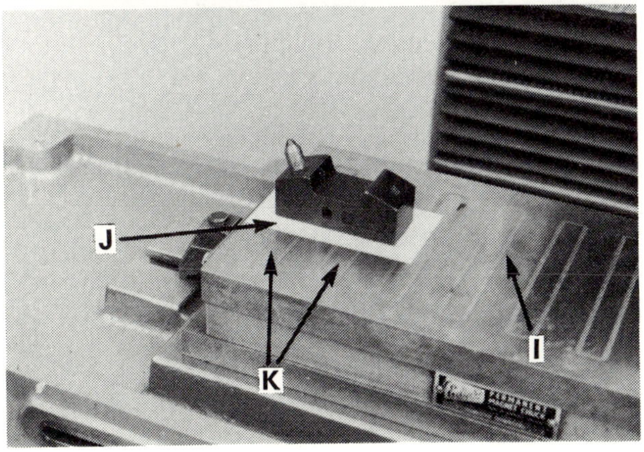

4. • Clean the surface of the magnetic chuck (I).
   • Place a piece of paper (J), slightly larger than the dresser base, on the left-hand end of the chuck.
   • Place the dresser on the paper being sure to straddle at least two magnetic inserts (K).
   • Energize the magnetic chuck.
   • Check that the diamond dresser is held securely by attempting to remove it from the chuck.

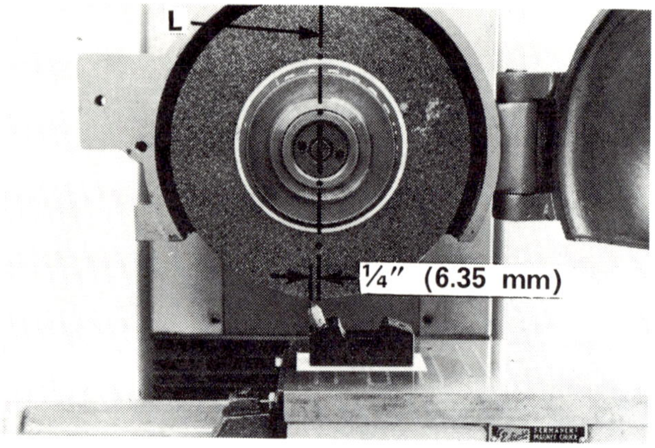

5. • Raise the wheel above the height of the diamond.
   • Move the table so that the diamond is offset approximately ¼″ (6.35 mm) to the left of the center line (L) of the wheel.
   (The diamond should be pointing in the same direction as the grinding wheel rotation.)

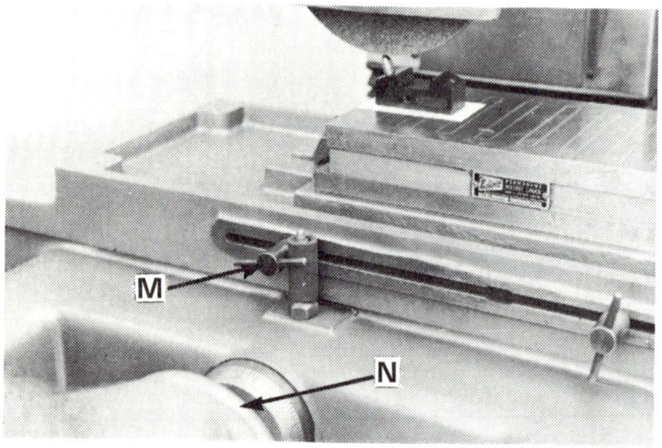

6. • Set the table stop (M) to keep the table and the diamond in position.
   • Start the grinding wheel.
   • Move the table with the crossfeed handwheel (N) until the diamond is under the high point on the face of the grinding wheel.

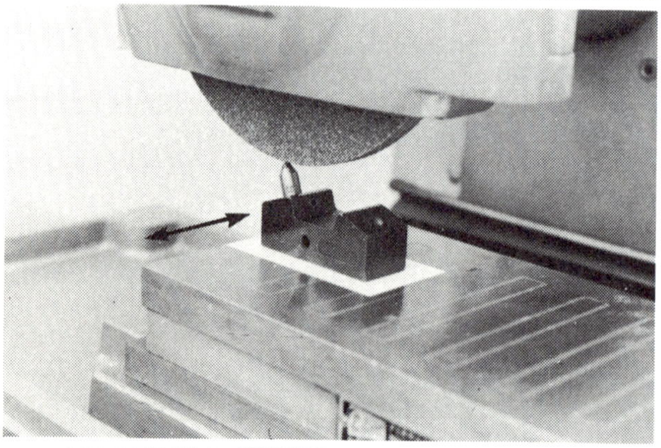

7. • Lower the wheelhead until the revolving wheel touches the diamond.
   • Use the crossfeed handwheel and move the diamond across the wheel face.
   • Lower the wheelhead .001-.002 (.02-.05 mm).
   • Pass the diamond across the wheel face in the direction of the arrows.
   • Dress until the wheelface is clean and flat.
   • Take a finish pass of .0005 (.01 mm) across the face of the grinding wheel.

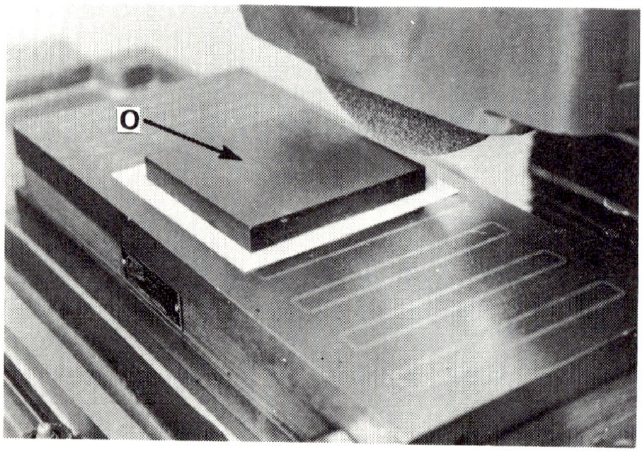

8. • Stop the grinding wheel.
   • Remove the diamond from the chuck.
   • Clean the table and mount the workpiece.
   • Take a light cut across the work surface.
   • If the wheel has been properly dressed, a good surface finish (O) should be produced on the work.

GRINDING A FLAT SURFACE

The most common operation on a surface grinder is grinding a flat or horizontal surface. To obtain the best results the correct type of wheel, properly dressed, should be used.

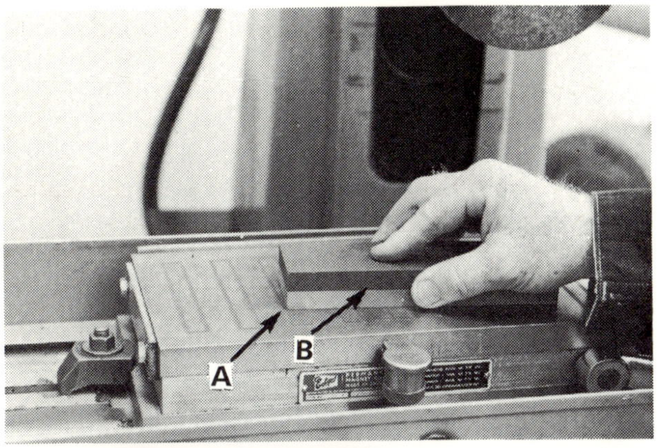

1. • **Dress the grinding wheel.**
   • **Clean the face of the magnetic chuck (A) with a cloth to remove any abrasive particles.**
   • **With a fine honing stone (B) remove all burrs from the face of the chuck.**
   • **Clean the face of the chuck.**

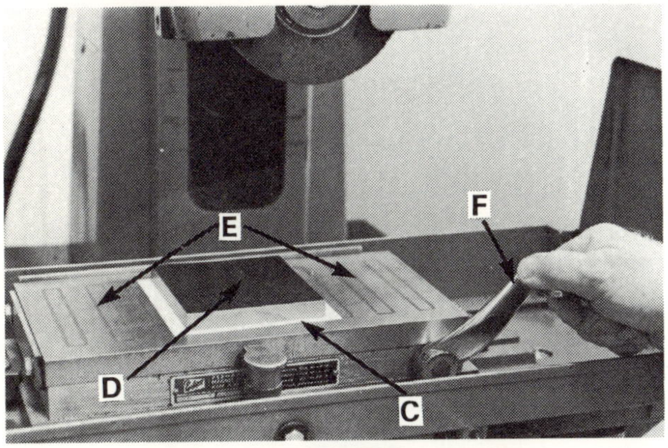

2. • **Remove all burrs from the workpiece.**
   • **Place a piece of paper (C), slightly larger than the workpiece, on the chuck face.**
   • **Place the workpiece (D) on top of the paper and straddle as many magnetic inserts (E) as possible.**
   • **Move the lever (F) to the ON position to energize the chuck.**
   • **Check that the work is held securely on the chuck by trying to remove it.**

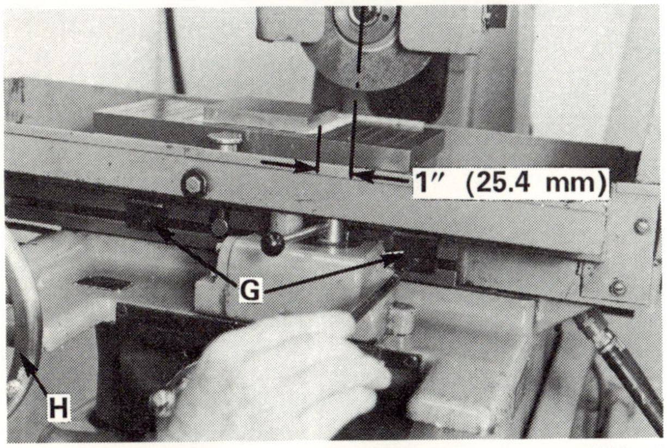

3. • **Set the table reverse dogs (G) so that the center of the grinding wheel clears each end of the workpiece by about 1" (25.4 mm).**
   • **With the table handwheel (H) move the table until the work is under the grinding wheel.**

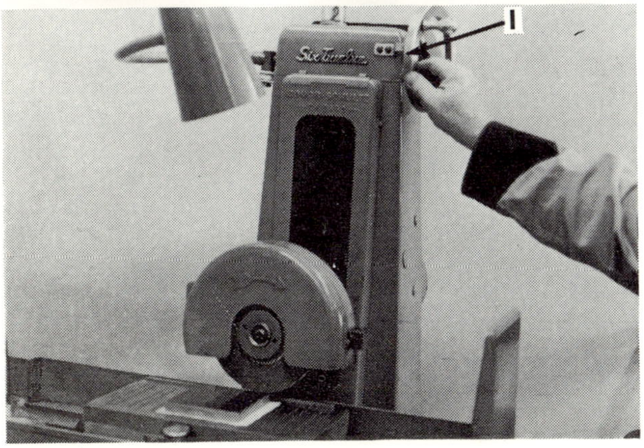

4. • **Turn the wheelfeed handwheel (I) until the grinding wheel is about 1/32" (.79 mm) above the work surface.**

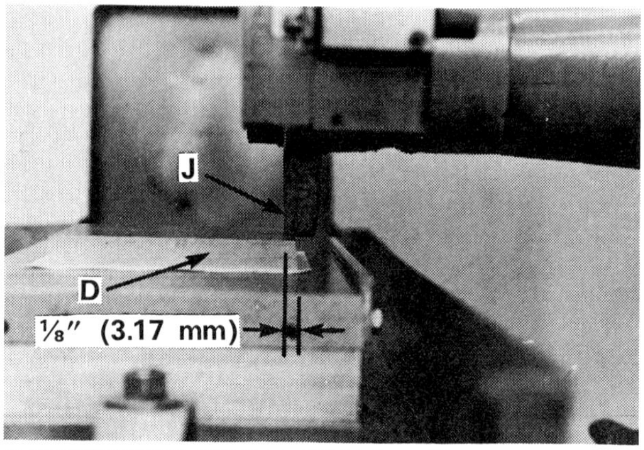

5. • Move the table with the crossfeed handwheel until the edge of the work (D) overlaps the edge of the grinding wheel (J) by about ⅛" (3.17 mm).

6. • Start the grinding wheel.
   • Carefully turn the wheelfeed handwheel (I) untill the grinding wheel just sparks the workpiece.
   • Turn the table crossfeed handle until the grinding wheel clears the edge of the work.

7. • With the wheelfeed handwheel, raise the grinding wheel .005 (.12 mm).
   • Start the table moving back and forth.
   • Feed across the entire surface of the work to locate any high spots.

   If any high spots are located, it will be necessary to again raise the wheel another .005 (.12 mm).

8. • Lower the table .002-.005 (.05-.13 mm).
   • Feed the table across using from .030-.050 a feed (.76-1.3 mm) per pass.
   • Rough grind the work until the surface just cleans up and check the surface finish.
   • Dress the wheel if necessary.
   • Finish grind the work using a cut of .0005-.001 (.01-.02 mm) deep.
   • Move the work clear of the wheel, shut off the machine and remove the work.

GRINDING THE EDGES OF A WORKPIECE SQUARE AND PARALLEL

When all surfaces of a workpiece are to be ground square and parallel, it is wise to follow a specific grinding sequence. Generally the larger flat surfaces are ground first and the edges are ground last.

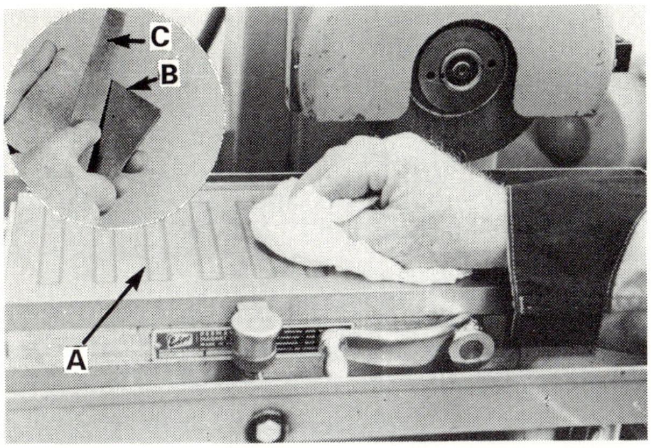

1. • Check the chuck for burrs and remove them with an oilstone.
   • Clean the surface of the chuck (A) with a cloth.
   • Remove all burrs from the workpiece (B) with a flat file (C).

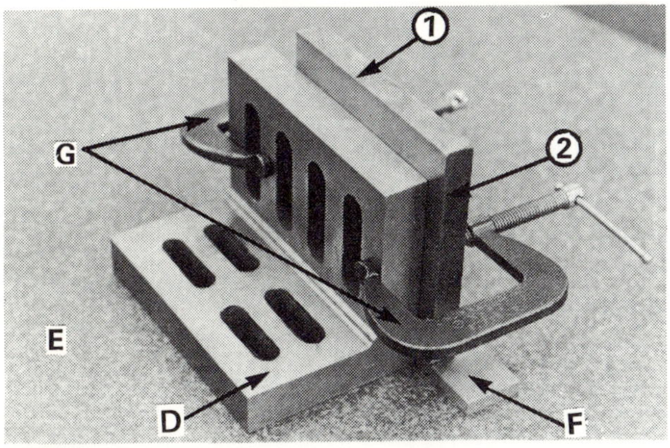

2. • Place a clean angle plate (D) on a surface plate (E).
   • Set the workpiece on a suitable parallel (F) with side #1 up and side #2 projecting about ½″ beyond the edge of the angle plate.
   • Fasten the work to the angle plate with suitable clamps (G).

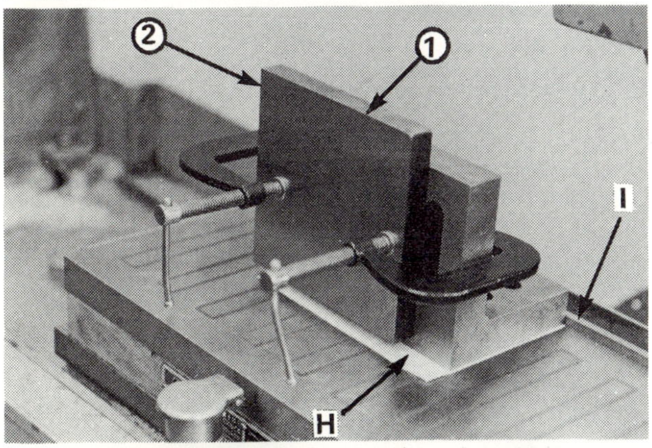

3. • Place a piece of paper (H), slightly larger than the base of the angle plate, on the magnetic chuck.
   • Place the angle plate and work on the paper.
   • Position the edge of the angle plate against the back stop bar (I) of the chuck.
   • Energize the magnetic chuck.
   • Check that the angle plate is held securely on the chuck by trying to remove it.

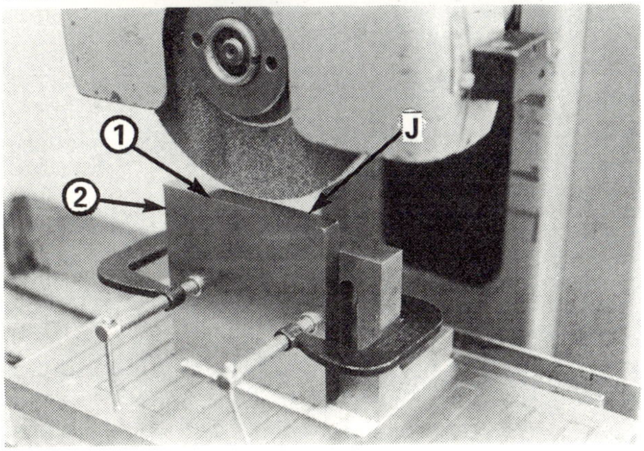

4. • Lower the grinding wheel to about 1/32″ (.79 mm) above side #1.
   • Adjust the table so that the work edge (J) overlaps the grinding wheel edge by ⅛″ (3.17 mm).
   • Start the grinding wheel.
   • Gradually lower the wheel until it just sparks the work.
   • Grind side #1 until the surface is clean and flat.

GRINDING THE EDGES OF A WORKPIECE SQUARE AND PARALLEL

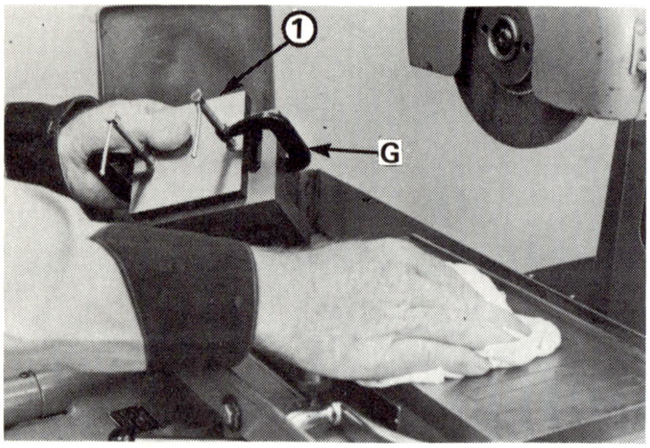

5. • Shut off the machine and raise the grinding wheel.
   • Remove the angle plate and work assembly from the magnetic chuck.
   • Clean the chuck face thoroughly.
   • Remove clamp (G) and place it vertically over side #1.
   DO NOT loosen the other clamp.
   • Tighten clamp (G) securely.

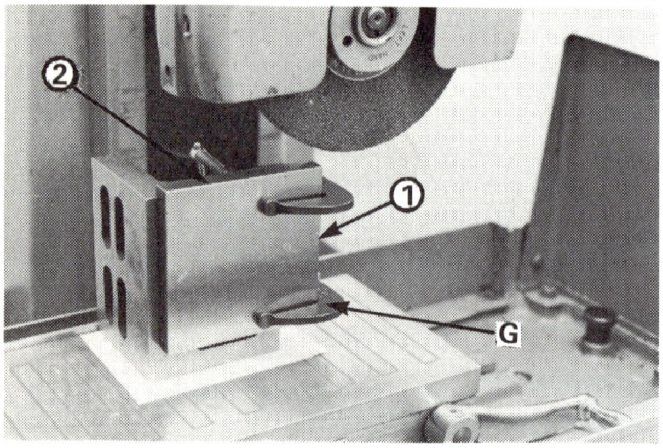

6. • Place clean paper on the chuck.
   • Place the edge of the angle plate on the paper with side #2 in the UP postiion.
   • Energize the magnetic chuck.
   • Re-position the upper clamp (G) to clear the top of side #2.
   • Grind side #2 until the surface is cleaned up.
   • Remove the work from the angle plate.

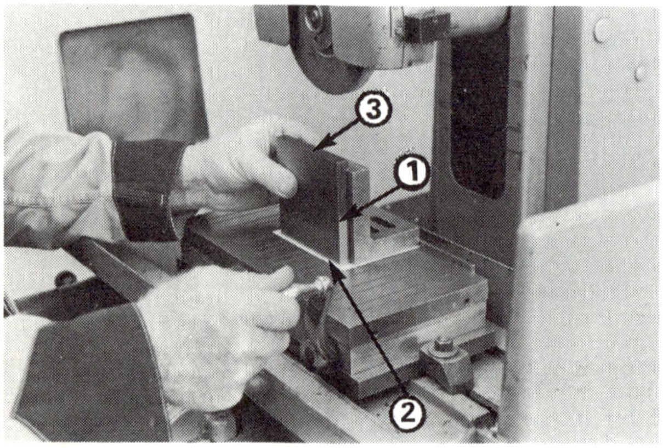

7. • Clean the chuck and mount the angle plate.
   • Remove all burrs from the work.
   • Place the work against the angle plate with #2 down on the paper.
   • Hold the work against the angle plate and energize the chuck.
   • Clamp the work to the angle plate.
   • Grind side #3 to the required size.
   • Remove the work from the angle plate.

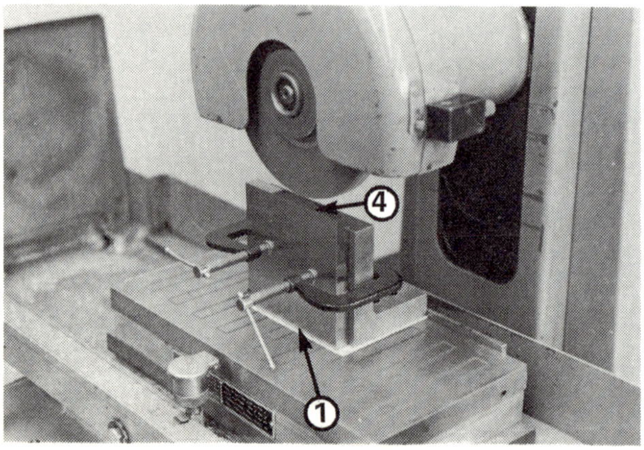

8. • Clean the chuck and mount the angle plate.
   • Remove all burrs from the work.
   • Place the work against the angle with side #1 down on the paper.
   • Hold the work against the angle plate and energize the chuck.
   • Clamp the work to the angle plate.
   • Grind side #4 to the required size.

Standard Tolerances for Press Fits

Light Press Fit

Dia. of Hole, in inches	Shaft Dia. Less Than Hole, in inches
Up to ½	+.0004 to +.0006
½ to 1	+.0005 to +.0010
1 to 2	+.00075 to +.0020
2 to 3	+.0015 to +.0030
3 to 4	+.0020 to +.0040
4 to 5	+.0020 to +.0045
5 to 6	+.0030 to +.0050

Heavy Press Fit

Dia. of Hole, in inches	Shaft Dia. Less Than Hole, in inches
Up to ½	+.005 to +.001
½ to 1	+.001 to +.003
1 to 2	+.002 to +.004
2 to 3	+.003 to +.006
3 to 4	+.005 to +.008
4 to 5	+.006 to +.010
5 to 6	+.008 to +.012

Standard Tolerances for Running Fits

Speeds up to 1000 r.p.m.

Dia. of Hole, in inches	Shaft Dia. Less Than Hole, in inches
Up to ½	—.0005 to —.0010
½ to 1	—.00075 to —.0015
1 to 2	—.0015 to —.0025
2 to 3	—.0020 to —.0025
3 to 4	—.0025 to —.0030
4 to 5	—.0030 to —.0035
5 to 6	—.0035 to —.0040

Speeds over 1000 r.p.m.

Dia. of Hole, in inches	Shaft Dia. Less Than Hole, in inches
Up to ½	—.0005 to —.0010
½ to 1	—.0010 to —.0020
1 to 2	—.0020 to —.0030
2 to 3	—.0025 to —.0035
3 to 4	—.0030 to —.0040
4 to 5	—.0035 to —.0045
5 to 6	—.0040 to —.0050

Standard Tolerances for Push Fits

Dia. of Hole, in inches	Shaft Dia. Less Than Hole, in inches
Up to ½	—.00025 to —.00075
½ to 1	—.0005 to —.0010
1 to 2	—.0005 to —.0015
2 to 3	—.0005 to —.0015
3 to 4	—.00075 to —.0020
4 to 5	—.00075 to —.0020
5 to 6	—.00075 to —.0020

Standard Tolerances for Sliding Fits

Dia. of Hole, in inches	Shaft Dia. Less Than Hole, in inches
Up to ½	—.005 to —.001
½ to 1	—.0075 to —.0015
1 to 2	—.0015 to —.0025
2 to 3	—.0020 to —.0030
3 to 4	—.0025 to —.0030
4 to 5	—.0025 to —.0035
5 to 6	—.0025 to —.0040

METRIC CONVERSION CHART

FRACTIONS			MILLIMETERS
		1/64	0.396
	1/32		0.793
		3/64	1.190
1/16			1.587
		5/64	1.984
	3/32		2.381
		7/64	2.778
1/8			3.175
		9/64	3.571
	5/32		3.968
		11/64	4.365
3/16			4.762
		13/64	5.159
	7/32		5.556
		15/64	5.953
1/4			6.350
		17/64	6.746
	9/32		7.143
		19/64	7.540
5/16			7.937
		21/64	8.334
	11/32		8.731
		23/64	9.128
3/8			9.525
		25/64	9.921
	13/32		10.318
		27/64	10.715
7/16			11.112
		29/64	11.509
	15/32		11.906
		31/64	12.303
1/2			12.700

FRACTIONS			MILLIMETERS
		33/64	13.096
	17/32		13.493
		35/64	13.890
9/16			14.287
		37/64	14.684
	19/32		15.081
		39/64	15.478
5/8			15.875
		41/64	16.271
	21/32		16.668
		43/64	17.065
11/16			17.462
		45/64	17.859
	23/32		18.256
		47/64	18.653
3/4			19.050
		49/64	19.446
	25/32		19.843
		51/64	20.240
13/16			20.637
		53/64	21.034
	27/32		21.431
		55/64	21.828
7/8			22.225
		57/64	22.621
	29/32		23.318
		59/64	23.415
15/16			23.812
		61/64	24.209
	31/32		24.606
		63/64	25.003
1.			25.400

FEED WITH COMPOUND REST AT 29°

T.P.I.	1	2	3	4	5	6	7	8	Finishing Cuts
8	.020	.020	.015	.015	.007	.005	.003	.002	.001
9	.020	.015	.015	.010	.007	.005	.003	.002	.001
10	.020	.015	.015	.010	.004	.003	.002	—	.001
11	.020	.015	.010	.007	.005	.003	.002	—	.001
13	.020	.015	.007	.005	.003	.002	—	—	.001
14	.015	.015	.010	.003	.002	.002	—	—	.001
16	.015	.010	.007	.005	.002	.002	—	—	.001
18	.015	.010	.005	.004	.002	—	—	—	.001